Barron's Regents Exams and Answers

Algebra II

Gary M. Rubinstein
B.A. Mathematics
Tufts University
M.S. Computer Science
University of Colorado

BARRON'S

Barron's Educational Series, Inc.

Published by Kaplan, Inc., d/b/a Barron's Educational Series
750 Third Avenue
New York, NY 10017
www.barronseduc.com

ISBN: 978-1-4380-0843-1
ISSN: 2472-9175

Printed in Canada

9 8 7 6 5 4 3 2 1

Kaplan, Inc., d/b/a Barron's Educational Series print books are available at
special quantity dipscounts to use for sales promotions, employee premiums,
or educational purposes. For more information or to purchase books, please
call the Simon & Schuster special sales department at 866-506-1949.

Contents

Regents Examinations, Answers, and Self-Analysis Charts   179

Preface

In 2009 New York State adopted the Common Core Standards in order to qualify for President Obama's Race To The Top initiative. The Common Core math is more difficult than the previous math curriculum, and the new state tests, including the Common Core Algebra II Regents, are more difficult as well.

Certain topics that had been in the Algebra II curriculum for decades have been removed for not being rigorous enough. Other topics have been added with the goal of making 21st-century American students more career and college ready than their predecessors.

The main topics that have been cut from the curriculum are permutations, combinations, Bernoulli trials, binomial expansion, and the majority of the trigonometric identities like the sine sum, cosine sum, sine difference, cosine difference, and the different double and half angle formulas. Other topics have been moved into earlier grades like the Law of Sines and the Law of Cosines. More than half the Trigonometry that had once been part of Algebra II is no longer part of the course.

Other topics have been added to fill the gaps left by those topics now considered obsolete. Primarily these are topics often taught in AP statistics as part of inferential statistics.

Aside from the change in topics, there is a change in the style of questions. There is a lot more need now for students to think more deeply about the topics as questions are intentionally phrased in a less straightforward way than they had been in the past.

Conquering the Algebra II test, something that was never an easy feat beforehand, has gotten that much more difficult and will require more test preparation than before. Getting this book is a great first step toward that goal. In addition to a review of all the topics that can appear on this test, there are nearly 1,000 practice questions of various levels of difficulty. The book can serve as a review or even as a way to learn the material for the first time. Teachers can also use this book to guide their pacing and to focus on the types of questions that are most likely to appear on the test while spending less focus on complicated aspects of the Common Core curriculum that are unlikely to be on the test.

The Common Core is part of a grand plan that is intended to propel our country to the top of the international rankings in math and reading. Good luck. We are all counting on you!

Gary Rubinstein
Math Teacher
2016

How to Use This Book

This book is designed to help you get the most out of your review for the new Regents exam in Algebra II (Common Core). Use this book to improve your understanding of the Algebra II topics and improve your grade.

TEST-TAKING TIPS

The first section in this book contains test-taking tips and strategies to help prepare you for the Algebra II Regents exam. This information is valuable, so be sure to read it carefully and refer to it during your study time. Remember: no single problem-solving strategy works for all problems—you should have a toolbox of strategies to pick from as you're facing unfamiliar or difficult problems on the test.

PRACTICE WITH KEY ALGEBRA II FACTS AND SKILLS

The second section in this book provides you with key Algebra II facts, useful skills, and practice problems with solutions. It provides you with a quick and easy way to refresh the skills you learned in class.

REGENTS EXAMS AND ANSWERS

The final section of the book contains actual Algebra II Regents. These exams and thorough answer explanations are probably the most useful tool for your review, as they let you know what's most important. By answering the questions on these exams, you will be able to identify your strengths and weaknesses and then concentrate on the areas in which you may need more study.

Remember, the answer explanations in this book are more than just simple solutions to the problems—they contain facts and explanations that are crucial to success in the Algebra II course and on the Regents exam. Careful review of these answers will increase your chances of doing well.

SELF-ANALYSIS CHARTS

Each of the Algebra II Regents exams ends with a Self-Analysis Chart. This chart will further help you identify weaknesses and direct your study efforts where needed. In addition, the chart classifies the questions on each exam into an organized set of topic groups. This feature will also help you to locate other questions on the same topic in the other Algebra II exam.

IMPORTANT TERMS TO KNOW

The terms that are listed in the glossary are the ones that have appeared most frequently on past Algebra II/Trigonometry and the most recent Algebra II (Common Core) exams. All terms and their definitions are conveniently organized for a quick reference.

Test-Taking Tips and Strategies

Knowing the material is only part of the battle in acing the new Algebra II Regents exam. Things like improper management of time, careless errors, and struggling with the calculator can cost valuable points. This section contains some test-taking strategies to help you perform your best on test day.

TIP 1

Time Management

SUGGESTIONS

- *Don't rush.* The Algebra II Regents exam is three hours long. While you are officially allowed to leave after 90 minutes, you really should stay until the end of the exam. Just as it wouldn't be wise to come to the test an hour late, it is almost as bad to leave a test an hour early.
- *Do the test twice.* The best way to protect against careless errors is to do the entire test twice and compare the answers you got the first time to the answers you got the second time. For any answers that don't agree, do a "tie breaker" third attempt. Redoing the test and comparing answers is much more effec-

tive than simply looking over your work. Students tend to miss careless errors when looking over their work. By redoing the questions, you are less likely to make the same mistake.

- *Bring a watch.* What will happen if the clock is broken? Without knowing how much time is left, you might rush and make careless errors. Yes, the proctor will probably write the time elapsed on the board and update it every so often, but it's better to be safe than sorry.

The TI-84 graphing calculator has a built in clock. Press the [MODE] to see it. If the time is not right, go to SET CLOCK and set it correctly. The TI-Nspire does not have a built-in clock.

TIP 2

Know How to Get Partial Credit

SUGGESTIONS

- *Know the structure of the exam.* The Algebra Regents exam has 37 questions. The first 24 of those questions are multiple-choice worth two points each. There is no partial credit if you make a mistake on one of those questions. Even the smallest careless error, like missing a negative sign, will result in no credit for that question. Parts II, III, and IV are free-response questions with no multiple-choice. Besides giving a numerical answer, you may be asked to explain your reasoning. Part II has eight free-response questions worth two points each. The smallest careless error will cause you to lose one point, which is half the value of the question. Part III has four free-response questions worth four points each. These questions generally have multiple parts. Part IV has one free-response question worth six points. This question will have multiple parts.

- *Explain your reasoning.* When a free-response question asks you to "Justify your answer," "Explain your answer," or "Explain how you determined your answer," the grader is expecting a few clearly written sentences. For these, you don't want to write too little since the grader needs to see that you understand why you did the different steps you did to solve the equation. You also don't want to write too much because if anything you write is not accurate, points can be deducted.

Here is an example followed by two solutions. The first would not get full credit, but the second would.

Example 1

Use algebra to solve for x in the equation $\frac{2}{3}x + 1 = 11$. Justify your steps.

Solution 1 (partial credit):

$\frac{2}{3}x + 1 = 11$ $-1 = -1$ $\frac{2}{3}x = 10$ $x = 15$	I used algebra to get the x by itself. The answer was $x = 15$.

Solution 2 (full credit):

$\frac{2}{3}x + 1 = 11$ $-1 = -1$ $\frac{2}{3}x = 10$ $\frac{3}{2} \cdot \frac{2}{3}x = \frac{3}{2} \cdot 10$ $1x = 15$ $x = 15$	I used the subtraction property of equality to eliminate the $+1$ from the left-hand side. Then to make it so the x had a 1 in front of it, I used the multiplication property of equality and multiplied both sides of the equation by the reciprocal of $\frac{2}{3}$, which is $\frac{3}{2}$. Then since $1 \cdot x = x$, the left-hand side of the equation just became x and the right-hand side became 15.

- *Computational errors vs. conceptual errors*

   In the Part III and Part IV questions, the graders are instructed to take off one point for a "computational error" but half credit for a "conceptual error." This is the difference between these two types of errors.

   If a four-point question was $x - 1 = 2$ and a student did it like this,

$$x - 1 = 2$$
$$\underline{+1 = +1}$$
$$x = 4$$

the student would lose one point out of 4 because there was one computational error since $2 + 1 = 3$ and not 4.

   Had the student done it like this,

$$x - 1 = 2$$
$$\underline{-1 = -1}$$
$$x = 1$$

the student would lose half credit, or 2 points, since this error was conceptual. The student thought that to eliminate the -1, he should subtract 1 from both sides of the equation.

   Either error might just be careless, but the conceptual error is the one that gets the harsher deduction.

TIP 3

Know Your Calculator

SUGGESTIONS

- *Which calculator should you use?* The two calculators used for this book are the TI-84 and the TI-Nspire. Both are very powerful. The TI-84 is somewhat easier to use for the functions needed for this test. The TI-Nspire has more features for courses in the future. The choice is up to you. This author prefers the TI-84 for the Algebra Regents. Graphing calculators come with manuals that are as thick as the book you are holding. There are also plenty of video tutorials online for learning how to use advanced features of the calculator. To become an expert user, watch the online tutorials or read the manual.
- *Clearing the memory.* You may be asked at the beginning of the test to clear the memory of your calculator. When practicing for the test, you should clear the memory too so you are practicing under test-taking conditions.

  This is how you clear the memory.

For the TI-84:

Press [2ND] and then [+] to get to the MEMORY menu. Then press [7] for Reset.

```
MEMORY
1:About
2:Mem Mgmt/Del…
3:Clear Entries
4:ClrAllLists
5:Archive
6:UnArchive
7▸Reset…
```

Use the arrows to go to [ALL] for All Memory. Then press [1].

```
RAM ARCHIVE ALL
1█All Memory...
```

Press [2] for Reset.

```
RESET MEMORY
1:No
2█Reset
Resetting ALL
will delete all
data, programs &
Apps from RAM &
Archive.
```

The calculator will be reset as if in brand new condition. The one setting that you may need to change is to turn the diagnostics on if you need to calculate the correlation coefficient.

For the TI-Nspire:

The TI-Nspire must be set to Press-to-Test mode when taking the Algebra Regents. Turn the calculator off by pressing [ctrl] and [home]. Press and hold [esc] and then press [home].

```
Press-to-Test
Prevent access to 3D graphing functionality
and pre-existing Scratchpad data, documents
and folders.
Angle Settings:  Degree ▶
Select additional restrictions:
☑ Limit geometry functions
☑ Disable function grab and move
      Enter Press-to-Test  Cancel
```

While in Press-to-Test mode, certain features will be deactivated. A small green light will blink on the calculator so a proctor can verify the calculator is in Press-to-Test mode.

To exit Press-to-Test mode, use a USB cable to connect the calculator to another TI-Nspire. Then from the home screen on the calculator in Press-to-Test mode, press [doc], [9], and select Exit Press-to-Test.

- *Use parentheses*
The calculator always uses the order of operations where multiplication and division happen before addition and subtraction. Sometimes, though, you may want the calculator to do the operations in a different order.

Suppose at the end of a quadratic equation, you have to round $x = \dfrac{-1+\sqrt{5}}{2}$ to the nearest hundredth. If you enter $(-)\,(1)\,(+)\,(2ND)\,(x^2)\,(5)\,(/)\,(2)$, it displays

which is not the correct answer.

One reason is that for the TI-84 there needs to be a closing parenthesis (or on the TI-Nspire, press [right arrow] to move out from under the radical sign) after the 5 in the square root symbol. Without it, it calculated $-1+\sqrt{\dfrac{5}{2}}$. More needs to be done, though, since

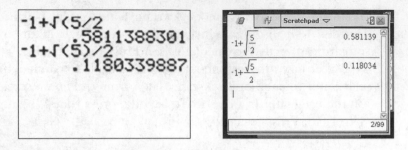

still is not correct. This is the solution to $-1+\sqrt{\dfrac{5}{2}}$.

To get this correct, there also needs to be parentheses around the entire numerator, $-1 + \sqrt{5}$.

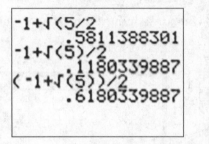

This is the correct answer.

On the TI-Nspire, fractions can also be done with [templates].

- *Using the ANS feature*

  The last number calculated with the calculator is stored in something called the ANS variable. This ANS variable will appear if you start an expression with a +, −, ×, or ÷. When an answer has a lot of digits in it, this saves time and is also more accurate.

  If for some step in a problem you need to calculate the decimal equivalent of $\frac{1}{7}$, it will look like this on the TI-84:

For the TI-Nspire, if you try the same thing, it leaves the answer as $\frac{1}{7}$. To get the decimal approximation, press [ctrl] and [enter] instead of just [enter].

Now if you want to multiply this by 3, just press [×], and the calculator will display "Ans*"; press [3] and [enter].

The ANS variable can also help you do calculations in stages. To calculate $x = \dfrac{-1 + \sqrt{5}}{2}$ without using so many parentheses as before, it can be done by first calculating $-1 + \sqrt{5}$ and then pressing [÷] and [2] and Ans will appear automatically.

The ANS variable can also be accessed by pressing [2ND] and [–] at the bottom right of the calculator. If after calculating the decimal equivalent of $\dfrac{1}{7}$ you wanted to subtract $\dfrac{1}{7}$ from 5, for the TI-84 press [5], [–], [2ND], [ANS], and [ENTER]. For the TI-Nspire press [5], [–], [ctrl], [ans], and [enter].

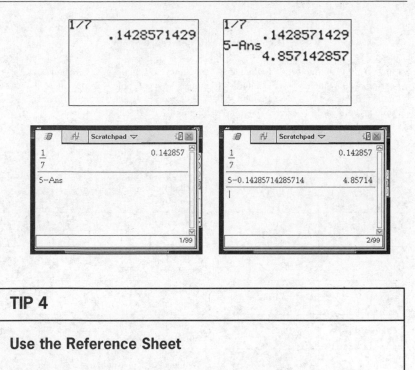

TIP 4
Use the Reference Sheet

SUGGESTIONS

- In the back of the Algebra II Regents booklet is a reference sheet that contains 17 conversion facts, such as inches to centimeters and quarts to pints, and also 17 formulas. Many of these conversion facts and formulas will not be needed for an individual test, but the quadratic formula and the arithmetic sequence formula are the two that will come in the handiest.

High School Math Reference Sheet

1 inch = 2.54 centimeters	1 kilometer = 0.62 mile	1 cup = 8 fluid ounces
1 meter = 39.37 inches	1 pound = 16 ounces	1 pint = 2 cups
1 mile = 5280 feet	1 pound = 0.454 kilogram	1 quart = 2 pints
1 mile = 1760 yards	1 kilogram = 2.2 pounds	1 gallon = 4 quarts
1 mile = 1.609 kilometers	1 ton = 2000 pounds	1 gallon = 3.785 liters
		1 liter = 0.264 gallon
		1 liter = 1000 cubic centimeters

Triangle	$A = \frac{1}{2}bh$	Pythagorean Theorem	$a^2 + b^2 = c^2$
Parallelogram	$A = bh$	Quadratic Formula	$x = \dfrac{-b \pm \sqrt{b^2 - 4ac}}{2a}$
Circle	$A = \pi r^2$	Arithmetic Sequence	$a_n = a_1 + (n-1)d$
Circle	$C = \pi d$ or $C = 2\pi r$	Geometric Sequence	$a_n = a_1 r^{n-1}$
General Prisms	$V = Bh$	Geometric Series	$S_n = \dfrac{a_1 - a_1 r^n}{1 - r}$ where $r \neq 1$
Cylinder	$V = \pi r^2 h$	Radians	1 radian = $\dfrac{180}{\pi}$ degrees
Sphere	$V = \frac{4}{3}\pi r^3$	Degrees	1 degree = $\dfrac{\pi}{180}$ radians
Cone	$V = \frac{1}{3}\pi r^2 h$	Exponential Growth/Decay	$A = A_0 e^{k(t - t_0)} + B_0$
Pyramid	$V = \frac{1}{3}Bh$		

HOW MANY POINTS DO YOU NEED TO PASS?

The Algebra II Regents exam is scored out of a possible 86 points. Unlike most tests given in the year by your teacher, the score is not then turned into a percent out of 86. Instead each test has a conversion sheet that varies from year to year. For the June 2016 test, the conversion sheet looked like this.

Raw Score	Scale Score	Raw Score	Scale Score	Raw Score	Scale Score
86	100	57	82	28	68
85	99	56	82	27	67
84	98	55	81	26	66
83	97	54	81	25	65
82	97	53	81	24	64
81	96	52	80	23	63
80	95	51	80	22	61
79	94	50	80	21	60
78	94	49	79	20	58
77	93	48	79	19	55
76	92	47	79	18	54
75	91	46	78	17	53
74	91	45	78	16	51
73	90	44	77	15	49
72	89	43	77	14	47
71	89	42	77	13	44
70	88	41	76	12	42
69	88	40	76	11	39
68	87	39	76	10	37
67	87	38	75	9	34
66	86	37	75	8	31
65	86	36	74	7	27
64	85	35	73	6	24
63	84	34	73	5	20
62	84	33	72	4	17
61	84	32	72	3	13
60	83	31	71	2	9
59	83	30	70	1	4
58	82	29	69	0	0

On this test, 30 points became a 70, 57 points became an 82, and 73 points became a 90. This means that for this examination a student who got 30 out of 86, which is just 35% of the possible points, would get a 73 on this exam. 57 out of 86 is 66%, but this scaled to an 86. 64 out of 86, however, is actually 85% and became an 85. So in the past there has been a curve on the exam for lower scores, though the scaling is not released until after the exam.

Some Key Algebra II Facts and Skills

1. POLYNOMIAL EXPRESSIONS AND EQUATIONS

1.1 POLYNOMIAL ARITHMETIC

A **polynomial** is an expression that combines numbers and variables raised to powers. $3x^2 + 5x + 1$ is an example of a polynomial.

The *terms* of this polynomial are $3x^2$, $5x$, and 1. Each term has a **coefficient**, which is the number in front of the term. In the term $3x^2$, the coefficient is 3.

Since the highest exponent in this polynomial is 2, we say that the **degree** of this polynomial is 2. Polynomials that have a degree of 2 are sometimes called **quadratic** polynomials. Polynomials that have a degree of 3 are sometimes called **cubic** polynomials. This polynomial has 3 terms. Polynomials with 3 terms are sometimes called **trinomials**. Polynomials with 2 terms are sometimes called **binomials**.

Adding and Subtracting Polynomials

When terms have the same variable and the same exponent, they are called **like terms**. Like terms can be easily added or subtracted.

$$5x^2 + 3x^2 = 8x^2$$
$$5x^2 - 3x^2 = 2x^2$$

Polynomials can be added or subtracted by combining their like terms.

$$(2x^2 + 5x - 3) + (3x^2 - 2x + 5) = 5x^2 + 3x + 2$$
$$(2x^2 + 5x - 3) - (3x^2 - 2x + 5) = -1x^2 + 7x - 8$$

Multiplying Polynomials

To multiply a polynomial by a constant, multiply that constant by each of the coefficients of each of the terms.

$$5(3x^2 + 2x - 1) = 15x^2 + 10x - 5$$

- To multiply a polynomial by another polynomial, each term in the polynomial on the left must be multiplied by each term in the polynomial on the right, and all those products should then be added together.

$$(2x + 3)(4x - 5) = 2x \cdot 4x + 2x(-5) + 3 \cdot 4x + 3(-5)$$
$$= 8x^2 - 10x + 12x - 15 = 8x^2 + 2x - 15$$

In this example, both polynomials were binomials. When multiplying two binomials, this process is often called **FOIL** since the four combinations represent the **F**irsts, **O**uters, **I**nners, and **L**asts.

Dividing Polynomials

Polynomial division is like long division.

Example
Find the quotient and remainder of $(6x^2 + 13x + 8) \div (2x + 1)$.

Solution:
- **Step 1**: Write the problem as you would for long division.

$$2x + 1 \overline{)6x^2 + 13x + 8}$$

Then, at each step, figure out what the first term of the divisor needs to be multiplied by.

- **Step 2**: Determine what $2x$ must be multiplied by to make it $6x^2$, and then multiply $2x + 1$ by that and subtract and bring down the next term.

$$
\begin{array}{r}
3x \\
2x+1{\overline{\smash{\big)}\,6x^2+13x+8}} \\
-\underline{(6x^2+3x)} \\
10x+8
\end{array}
$$

- **Step 3**: Determine what $2x$ must be multiplied by to make it $10x$, and then multiply $2x + 1$ by that and subtract. There will be no other terms to bring down.

$$
\begin{array}{r}
3x+5 \\
2x+1{\overline{\smash{\big)}\,6x^2+13x+8}} \\
-\underline{(6x^2+3x)} \\
10x+8 \\
-\underline{10x+5} \\
3
\end{array}
$$

- **Step 4**: The solution is $3x + 5$ and the remainder is 3. You can write $3x + 5$ R3 or as a fractional remainder $3x + 5 + \dfrac{3}{2x+1}$.

1.2 FACTORING POLYNOMIALS

Greatest Common Factor Factoring
If all the terms of a polynomial have a common factor, that factor can be "factored out."

Example
Factor $6x^3 + 15x^2 + 21x$.

Solution:
Since each coefficient is divisible by 3 and each variable part is divisible by x, the common factor is $3x$. When it is factored out, it becomes $3x(2x^2 + 5x + 7)$.

Difference of Perfect Square Factoring
A polynomial of the form $x^2 - a^2$ can be factored into $(x - a)(x + a)$.

Example
Factor $x^2 - 9$.

Solution:
Since this can be written as $x^2 - 3^2$, it can be factored into

$$(x - 3)(x + 3)$$

Perfect Square Trinomial Factoring
A polynomial of the form $x^2 + 2ax + a^2$ can be factored into

$$(x + a)(x + a) = (x + a)^2$$

Example
Factor $x^2 + 10x + 25$.

Solution:
Since this can be written as $x^2 + 2 \cdot 5 \cdot x + 5^2$, it can be factored into $(x + 5)^2$. In general, if the square of half the coefficient of the x is equal to the constant term, this factoring pattern can be used. Since $\left(\dfrac{10}{2}\right)^2 = 25$, this factoring pattern applies.

General Trinomial Factoring

A polynomial of the form $x^2 + bx + c$ can be factored into

$$(x + m)(x + n)$$

if m and n can be found so that $m + n = b$ and $m \cdot n = c$.

Example

Factor $x^2 + 5x + 6$.

Solution:

Since there are two numbers 2 and 3 such that $2 + 3 = 5$ and $2 \cdot 3 = 6$, this can be factored into $(x + 2)(x + 3)$.

Factor by Grouping

Certain cubic expressions can be factored by a two-step process called **grouping**.

Example

Factor $2x^3 + 4x^2 + 6x + 12$.

Solution:

- **Step 1:** Find a common factor for the first two terms, and factor it out of them: $2x^2(x + 2) + 6x + 12$

- **Step 2:** Find a common factor for the last two terms, and factor it out of them: $2x^2(x + 2) + 6(x + 2)$

- **Step 3:** If these two parts have a common factor, as they do in this case with $(x + 2)$, factor that out of both terms:

$$(x + 2)(2x^2 + 6)$$

Notice that $2x^2 + 6$ can be further factored as $2(x^2 + 3)$. The final answer is $2(x + 2)(x^2 + 3)$.

1.3 THE REMAINDER THEOREM AND THE FACTOR THEOREM

There is a short cut for finding the remainder when a polynomial is divided by a binomial of the form $x - a$.

MATH FACTS

The remainder when a polynomial is divided by $x - a$ will always be the same as the value of the polynomial when $x = a$ is substituted into it.

Example
What is the remainder of $(x^2 + 5x - 8) \div (x - 3)$?

Solution:
According to the Remainder Theorem, the remainder will be equal to $3^2 + 5 \cdot 3 - 8 = 9 + 15 - 8 = 16$.

MATH FACTS

There is a shortcut for determining if $x - a$ is a factor of another polynomial. If the value of the polynomial is zero when $x = a$ is substituted into it, then $x - a$ is a factor.

Example
Is $x - 3$ a factor of $x^3 + 2x^2 - 12x - 9$?

Solution:
Since $3^3 + 2 \cdot 3^2 - 12 \cdot 3 - 9 = 27 + 18 - 36 - 9 = 0$, $x - 3$ is a factor of $x^3 + 2x^2 - 12x - 9$.

1.4 POLYNOMIAL EQUATIONS

A **polynomial equation** has a polynomial on either one or both sides of the equal sign. The simplest type of polynomial equation has a zero on one side of the equal sign and a polynomial that can be factored on the other side.

Example
Solve for all values of x if $x^2 + 7x + 10 = 0$.

Solution:

$$x^2 + 7x + 10 = 0$$
$$(x + 2)(x + 5) = 0$$

For a product to be zero, at least one of the factors must be zero so either $x + 2 = 0$ or $x + 5 = 0$.

$$\begin{array}{ccc} x + 2 = 0 & \text{or} & x + 5 = 0 \\ \underline{-2 = -2} & & \underline{-5 = -5} \\ x = -2 & \text{or} & x = -5 \end{array}$$

The solution set is $\{-2, -5\}$.

Solving Quadratic Equations When the Polynomial Cannot Be Factored

An equation like $x^2 + 6x + 7 = 0$ cannot be solved like the previous example since the polynomial cannot be factored. When this happens, there are two methods for solving it.

Method 1: Completing the Square

- **Step 1:** Eliminate the constant from the right side by subtracting 7 from both sides.

$$\begin{array}{r} x^2 + 6x + 7 = 0 \\ \underline{-7 = -7} \\ x^2 + 6x = -7 \end{array}$$

- **Step 2:** "Complete" the square by adding the square of half the coefficient on the x to both sides of the equation. For this example, add $\left(\dfrac{6}{2}\right)^2 = 9$ to both sides.

$$x^2 + 6x = -7$$
$$+9 = +9$$
$$\overline{x^2 + 6x + 9 = 2}$$

- **Step 3:** The left-side polynomial is now a perfect square polynomial, which can be factored to $(x + 3)^2$. Take the square root of both sides and eliminate the constant from the left-hand side.

$$x^2 + 6x + 9 = 2$$
$$(x + 3)^2 = 2$$
$$\sqrt{(x + 3)^2} = \pm\sqrt{2}$$
$$x + 3 = \pm\sqrt{2}$$
$$-3 = -3$$
$$\overline{x = -3 \pm \sqrt{2}}$$

Method 2: The Quadratic Formula

To solve equations of the form $ax^2 + bx + c = 0$, use the formula $x = \dfrac{-b \pm \sqrt{b^2 - 4ac}}{2a}$. For this example, $a = 1$, $b = 6$, and $c = 7$.

$$x = \frac{-6 \pm \sqrt{6^2 - 4 \cdot 1 \cdot 7}}{2 \cdot 1}$$
$$= \frac{-6 \pm \sqrt{36 - 28}}{2}$$
$$= \frac{-6 \pm \sqrt{8}}{2}$$
$$= \frac{-6 \pm 2\sqrt{2}}{2}$$
$$= -3 \pm \sqrt{2}$$

1.5 GRAPHS OF EQUATIC EXPRESSIONS

The graph of $y = ax^2 + bx + c$ will be a parabola. If $a > 0$, the parabola will open "upward" like the letter u. If $a < 0$, the parabola will open "downward" like the letter n.

The graph of $y = x^2 + 6x + 8$ looks like this:

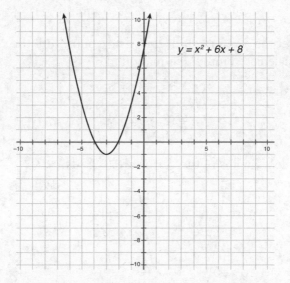

- Four important points on the parabola are the vertex, the y-intercept, and the x-intercepts. There can be 0, 1, or 2 x-intercepts.

The y-intercept is at $(0, c)$, which is $(0, 8)$ for $y = x^2 + 6x + 8$.

The x-intercept(s) can be found by solving, if possible, the equation $0 = x^2 + 6x + 8$. This might require factoring, completing the square, or using the quadratic formula.

$$0 = x^2 + 6x + 8$$
$$0 = (x + 2)(x + 4)$$

$$
\begin{array}{ccc}
x + 2 = 0 & \text{or} & x + 4 = 0 \\
\underline{-2 = -2} & & \underline{-4 = -4} \\
x = -2 & \text{or} & x = -4
\end{array}
$$

The x-intercepts are $(-2, 0)$ and $(-4, 0)$.

The vertex is the low point if the parabola looks like a u and the high point if it looks like an n. The x-coordinate of the vertex is always $-\dfrac{b}{2a}$. For this example, that is

$$-\frac{6}{2 \cdot 1} = -\frac{6}{2} = -3$$

To get the y-intercept of the vertex, substitute the x-value you just calculated into the equation.

$$y = x^2 + 6x + 8$$
$$y = (-3)^2 + 6(-3) + 8$$
$$y = 9 - 18 + 8$$
$$y = -1$$

The coordinates of the vertex are $(-3, -1)$.

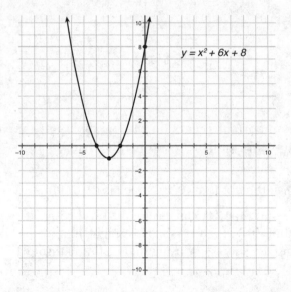

$y = x^2 + 6x + 8$

- If an equation has the form $y = a(x - h)^2 + k$, the graph will also be a parabola. If $a > 0$, it looks like a u. Otherwise, it looks like an n. This form is called *vertex form* since the numbers in the place of h and k will be the vertex of the parabola.

Example

What is the vertex of the graph of $y = 1(x + 3)^2 - 1$?

Solution:

Since this can be written as $y = 1(x - (-3))^2 + (-1)$, the h is -3 and the k is -1 so the vertex is at $(-3, -1)$.

Focus and Directrix of a Parabola

- Every parabola also has associated with it an invisible horizontal line called the **directrix** and an invisible point called the **focus**.

If the equation is $y = ax^2 + bx + c$ or $y = a(x - h)^2 + k$, the focus is $\frac{1}{4a}$ units above the vertex, and the directrix is $\frac{1}{4a}$ units below the vertex.

For $y = x^2 + 6x + 8$, the vertex is $(-3, -1)$ and $\frac{1}{4a} = \frac{1}{4 \cdot 1} = \frac{1}{4}$ so the focus is $\frac{1}{4}$ units above the vertex at $\left(-3, -\frac{3}{4}\right)$ and the directrix is $y = -1\frac{1}{4}$.

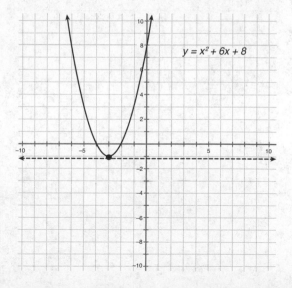

$y = x^2 + 6x + 8$

Graphs of Higher-Degree Polynomials

For degree greater than 2, the graph of the polynomial resembles a snake with the number of turns usually one less than the degree.

- The graph of the cubic $y = (x - 1)(x + 1)(x - 2)$ has three x-intercepts because $0 = (x - 1)(x + 1)(x - 2)$ has three solutions.

$$
\begin{array}{ccccc}
x - 1 = 0 & \text{or} & x + 1 = 0 & \text{or} & x - 2 = 0 \\
x = 1 & \text{or} & x = -1 & \text{or} & x = 2
\end{array}
$$

The three x-intercepts are $(1, 0)$, $(-1, 0)$, and $(2, 0)$.

The graph of $y = (x - 1)(x + 1)(x - 2)$ looks like this

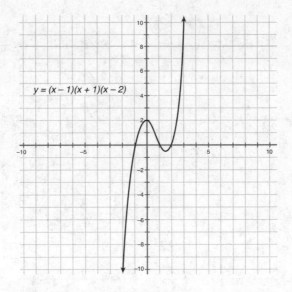

A **double root** happens when two of the x-intercepts are located at the same point.

In the equation $y=(x-2)(x-2)(x+1)$, the three x-intercepts are $(2, 0)$, $(2, 0)$, and $(-1, 0)$. In the graph, the curve "bounces off" the x-axis at $(2, 0)$, the location of the double root.

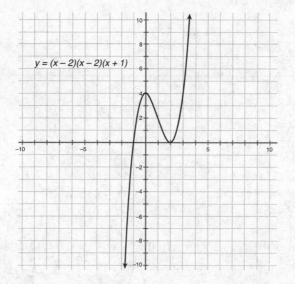

Practice Exercises

1. What is $(5x^2 - 3x + 8) - (2x^2 + 4x - 2)$?
   (1) $3x^2 - 7x + 10$
   (2) $3x^2 + x + 10$
   (3) $3x^2 - 7x + 6$
   (4) $3x^2 + x + 6$

2. Which shows $x^2 - 12x + 36$ factored?
   (1) $(x + 6)^2$
   (2) $(x + 6)(x - 6)$
   (3) $(x - 1)(x - 36)$
   (4) $(x - 6)^2$

3. What is the remainder when $x^3 + x^2 - 9x - 5$ is divided by $x - 3$?
   (1) 4
   (2) 5
   (3) 6
   (4) 7

4. Which of the following is a factor of $x^3 + 3x^2 - 10x - 24$?
   (1) $(x - 1)$
   (2) $(x - 2)$
   (3) $(x - 3)$
   (4) $(x - 4)$

5. What are the solutions to $x^2 - 6x + 4 = 0$?
   (1) $x = 5 \pm \sqrt{3}$
   (2) $x = 3 \pm \sqrt{2}$
   (3) $x = 3 \pm \sqrt{5}$
   (4) $x = 2 \pm \sqrt{5}$

6. Which is an equation of a parabola with its vertex at $(3, -1)$?
   (1) $y = (x - 3)^2 + 1$ (3) $y = (x + 3)^2 + 1$
   (2) $y = (x - 3)^2 - 1$ (4) $y = (x + 3)^2 - 1$

7. What is the vertex of the parabola whose equation is
   $y = x^2 + 4x - 12$?
   (1) $(-2, -16)$ (3) $(2, 0)$
   (2) $(-2, -24)$ (4) $(2, -16)$

8. What is the equation of the directrix and the coordinates of the
   focus of the parabola $y = 3x^2$?

   (1) $y = -12$ and $(0, 12)$ (3) $y = -\dfrac{1}{12}$ and $\left(0, \dfrac{1}{12}\right)$

   (2) $y = -\dfrac{1}{4}$ and $\left(0, \dfrac{1}{4}\right)$ (4) $y = \dfrac{1}{12}$ and $\left(0, -\dfrac{1}{12}\right)$

9. What are the x-intercepts and y-intercept of
   $y = (x - 2)(x + 4)(x + 5)$?
   (1) $(-2, 0), (4, 0), (5, 0), (0, -40)$
   (2) $(-2, 0), (4, 0), (5, 0), (0, 40)$
   (3) $(2, 0), (-4, 0), (-5, 0), (0, -40)$
   (4) $(2, 0), (-4, 0), (-5, 0), (0, 40)$

10. If a and b are positive integers, which could be the graph of
$y = (x - b)(x^2 + 2ax + a^2)$?

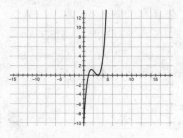

(1)

(3)

(2)

(4)

Solutions

1. $(5x^2 - 3x + 8) - (2x^2 + 4x - 2)$
    $5x^2 - 3x + 8 - 2x^2 - 4x + 2$
    $3x^2 - 7x + 10$

    The correct choice is (**1**).

2. Because the square of half the b is equal to the constant term,

    c, $\left(\dfrac{-12}{2}\right)^2 = 36$, this is a perfect square trinomial and can be

    factored into $\left(x + \dfrac{b}{2}\right)^2 = \left(x + \dfrac{-12}{2}\right)^2 = (x - 6)^2$.

    The correct choice is (**4**).

3. The Remainder Theorem says that you can get the remainder when dividing by $x - a$ by substituting a for x into the expression you are dividing into.

    $$3^3 + 3^2 - 9(3) - 5 = 27 + 9 - 27 - 5 = 4$$

    The correct choice is (**1**).

4. To check if $(x - a)$ is a factor of a polynomial, substitute a for x into the polynomial, and if it evaluates to 0, $(x - a)$ is a factor. Testing all the choices, this only happens for choice (3).

    $$3^3 + 3 \cdot 3^2 - 10(3) - 24 = 27 + 27 - 30 - 24 = 0$$

    The correct choice is (**3**).

5. Using the quadratic formula,

$$x = \frac{-(-6) \pm \sqrt{(-6)^2 - 4(1)(4)}}{2(1)}$$

$$= \frac{6 \pm \sqrt{36 - 16}}{2}$$

$$= \frac{6 \pm \sqrt{20}}{2}$$

$$= \frac{6 \pm 2\sqrt{5}}{2}$$

$$= 3 \pm \sqrt{5}$$

The correct choice is **(3)**.

6. When the equation is in vertex form, $y = (x - h)^2 + k$, the vertex of the parabola will be (h, k). If the vertex is $(3, -1)$, the equation is $y = (x - 3)^2 + (-1) = (x - 3)^2 - 1$.

The correct choice is **(2)**.

7. The x-coordinate of the vertex can be found with the formula

$$x = \frac{-b}{2a} = \frac{-4}{2(1)} = -2$$

The y-coordinate can be found by substituting the x-coordinate into the equation and solving for y.

$$y = (-2)^2 + 4(-2) - 12 = 4 - 8 - 12 = -16$$

The vertex is $(-2, -16)$.

The correct choice is **(1)**.

8. The focus is $\dfrac{1}{4a}$ units above the vertex. When the equation for a parabola is of the form $y = ax^2$, the vertex is $(0, 0)$. Since $a = 3$, $\dfrac{1}{4a} = \dfrac{1}{4(3)} = \dfrac{1}{12}$. So the focus is $\left(0, 0 + \dfrac{1}{12}\right) = \left(0, \dfrac{1}{12}\right)$.

The directrix is a horizontal line $\dfrac{1}{4a}$ units below the vertex. So the directrix is $y = 0 - \dfrac{1}{12} = -\dfrac{1}{12}$.

The correct choice is **(3)**.

9. To find the y-intercept, substitute 0 into the equation

$$y = (0 - 2)(0 + 4)(0 + 5) = (-2)(4)(5) = -40$$

so the y-intercept is $(0, -40)$.

To find the x-intercepts, find the solutions to the equation

$$0 = (x - 2)(x + 4)(x + 5)$$

To get a product of zero, one of the factors must be 0 so either $x - 2 = 0$, $x + 4 = 0$, or $x + 5 = 0$. The solutions to these three equations are 2, –4, and –5 so the x-intercepts are $(2, 0)$, $(-4, 0)$, and $(-5, 0)$.

The correct choice is **(3)**.

10. The second factor is a perfect square trinomial so this can be factored to $y = (x - b)(x + a)(x + a) = (x - b)(x + a)^2$. This means that there is a root at $x = b$ to the right of the y-axis, and a double root at $x = -a$ to the left of the y-axis. At a double root, the curve bounces off the x-axis. This happens in choice (1).

The correct choice is **(1)**.

2. RATIONAL EXPRESSIONS AND EQUATIONS

2.1 ARITHMETIC WITH RATIONAL EXPRESSIONS

- A **rational expression** involves a fraction that has a polynomial in its denominator.

  $\dfrac{1}{x^2+5x+6}$ is a rational expression.

  A rational expression can have a polynomial in both the denominator and the numerator, such as the rational expression $\dfrac{2x+6}{x^2+5x+6}$.

Reducing Rational Expressions

To reduce a rational expression, factor the numerator and the denominator and "cancel out" any common factors.

Example

Reduce the expression $\dfrac{2x+6}{x^2+5x+6}$.

Solution:

$$\frac{2x+6}{x^2+5x+6} = \frac{2\cancel{(x+3)}}{(x+2)\cancel{(x+3)}} = \frac{2}{x+2}$$

A Warning About 'Canceling Out'

When the numerator has two or more terms separated by a + or a −, you can only 'cancel out' something in the denominator if each of the terms in the numerator shares the common factor.

So

$$\frac{2(x+5)+5(x+5)}{(x+2)(x+5)} = \frac{2\cancel{(x+5)}+5\cancel{(x+5)}}{(x+2)\cancel{(x+5)}} = \frac{2+5}{x+2} = \frac{7}{x+2}$$

But you cannot do

$$\frac{2(x+3)+5}{(x+2)(x+3)} = \frac{2\cancel{(x+3)}+5}{(x+2)\cancel{(x+3)}} = \frac{2+5}{x+2} = \frac{7}{x+2}$$

since the 5 did not get divided by the $(x + 3)$.

Multiplying Rational Expressions

To multiply two rational expressions, first factor all polynomials in the numerators and denominators. You can then cancel out any factors in either of the denominators with any matching factors in any of the numerators.

Example

Multiply $\dfrac{2x+6}{5x-10} \cdot \dfrac{x^2-4}{x^2+5x+6} = \dfrac{2\cancel{(x+3)}}{5\cancel{(x-2)}} \cdot \dfrac{\cancel{(x-2)}(x+2)}{\cancel{(x+2)}\cancel{(x+3)}} = \dfrac{2}{5}$

Dividing Rational Expressions

Divide by multiplying the expression on the left by the reciprocal of the expression on the right. If the division is represented as a fraction, multiply the numerator by the reciprocal of the denominator.

Example

$$\frac{5}{x+3} \div \frac{2}{x^2+5x+6} = \frac{5}{x+3} \cdot \frac{x^2+5x+6}{2}$$

$$= \frac{5}{\cancel{x+3}} \cdot \frac{(x+2)\cancel{(x+3)}}{2}$$

$$= \frac{5(x+2)}{2}$$

Adding Rational Expressions

If two rational expressions have the same denominator, their sum has that same denominator, and the numerator is the sum of the two numerators.

Example

$$\frac{5x}{x+3}+\frac{2}{x+3}=\frac{5x+2}{x+3}$$

- If two rational expressions have different denominators that have no common factor, the numerator and denominator of each rational expression must first be multiplied by the denominator of the other rational expression. The new rational expressions will have a common denominator and can then be added.

Example

What is $\dfrac{5}{x+2}+\dfrac{3}{x+5}$?

Solution:

$$\frac{5}{x+2}+\frac{3}{x+5}=\frac{5(x+5)}{(x+2)(x+5)}+\frac{3(x+2)}{(x+2)(x+5)}=\frac{5(x+5)+3(x+2)}{(x+2)(x+5)}$$
$$=\frac{5x+25+3x+6}{(x+2)(x+5)}=\frac{8x+31}{(x+2)(x+5)}$$

- If two rational expressions have different denominators that *do* have a common factor, the common denominator will have each of the factors of each of the denominators in it, but just one of the common factors.

Example

What is $\dfrac{5}{(x+2)(x+3)}+\dfrac{3}{(x+2)(x+5)}$?

Solution:

Because both denominators have a factor of $(x + 2)$, the common denominator will be $(x + 2)(x + 3)(x + 5)$.

$$\frac{5}{(x+2)(x+3)} + \frac{3}{(x+2)(x+5)}$$

$$= \frac{5(x+5)}{(x+2)(x+3)(x+5)} + \frac{3(x+3)}{(x+2)(x+3)(x+5)}$$

$$= \frac{5(x+5)+3(x+3)}{(x+2)(x+3)(x+5)} = \frac{5x+25+3x+9}{(x+2)(x+3)(x+5)}$$

$$= \frac{8x+34}{(x+2)(x+3)(x+5)}$$

Subtracting Rational Expressions

Subtracting is just like adding, but you need to be more careful when subtracting since people sometimes improperly distribute the negative sign through the right-hand expression.

Example

What is $\dfrac{5}{x+2} - \dfrac{3}{x+5}$?

Solution:

$$\frac{5}{x+2} - \frac{3}{x+5} = \frac{5(x+5)}{(x+2)(x+5)} - \frac{3(x+2)}{(x+2)(x+5)} = \frac{5(x+5)-3(x+2)}{(x+2)(x+5)}$$

$$= \frac{5x+25-3x-6}{(x+2)(x+5)} = \frac{2x+19}{(x+2)(x+5)}$$

A common error is to have a +6 instead of a –6 in the second to last step because of not distributing the –3 through.

2.2 RATIONAL EQUATIONS

A **rational equation** involves one or more rational expressions, like $\dfrac{2}{x+2} + \dfrac{3}{x+5} = \dfrac{7x+10}{(x+2)(x+5)}$.

- **Step 1**: Combine the two rational expressions using the methods from the previous section.

$$\frac{2}{x+2} + \frac{3}{x+5} = \frac{7x+10}{(x+2)(x+5)}$$

$$\frac{2(x+5)}{(x+2)(x+5)} + \frac{3(x+2)}{(x+2)(x+5)} = \frac{7x+10}{(x+2)(x+5)}$$

$$\frac{2(x+5) + 3(x+2)}{(x+2)(x+5)} = \frac{7x+10}{(x+2)(x+5)}$$

$$\frac{2x+10+3x+6}{(x+2)(x+5)} = \frac{7x+10}{(x+2)(x+5)}$$

$$\frac{5x+16}{(x+2)(x+5)} = \frac{7x+10}{(x+2)(x+5)}$$

- **Step 2**: If the denominators are different, cross multiply. If the denominators are the same, as they are in this example, they can be ignored.

$$5x+16 = 7x+10$$
$$-5x = -5x$$
$$16 = 2x+10$$
$$-10 = -10$$
$$\overline{6 = 2x}$$
$$\frac{6}{2} = \frac{2x}{2}$$
$$3 = x$$

- **Step 3**: Check your answer. Sometimes this process produces "fake" answers so the only way to know whether or not to reject any potential solutions is to substitute back into the *original* equation and see if the solution makes that equation true.

Checking $x = 3$:

$$\frac{2}{3+2} + \frac{3}{3+5} = \frac{7 \cdot 3 + 10}{(3+2)(3+5)}$$

$$\frac{2}{5} + \frac{3}{8} = \frac{31}{40}$$

$$\frac{16}{40} + \frac{15}{40} = \frac{31}{40}$$

$$\frac{31}{40} = \frac{31}{40}$$

Since this is true, $x = 3$ is the solution to the original equation.

2.3 GRAPHS OF RATIONAL FUNCTIONS

The graph of a rational function like $y = \dfrac{2x-5}{x-3}$ has some properties that graphs of polynomial functions do not.

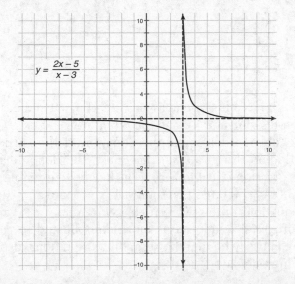

An asymptote of a graph is a line that is not part of the graph but functions like a wall that the graph gets closer and closer to without touching. The dotted lines indicate the location of these invisible asymptotes of the graph.

The graph above has a horizontal asymptote of $y = 2$ and a vertical asymptote of $x = 3$.

The graph of a rational function can have more than one vertical asymptote. Below is the graph of $y = \dfrac{3}{(x-3)(x+1)}$.

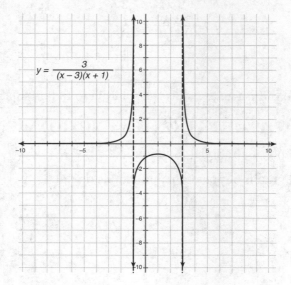

This graph has a horizontal asymptote of $y = 0$ and has two vertical asymptotes, one at $x = 3$ and the other at $x = -1$.

It is possible to determine the equation of the vertical asymptotes by finding the x-value(s) that make the denominator of the rational expression equal to zero. For the above example, the equation $(x - 3)(x + 1) = 0$ has two solutions, $x = 3$ and $x = -1$, which are also the equations of the two vertical asymptotes.

The horizontal asymptote is more complicated to calculate by hand but can easily be found by graphing the rational expression on the graphing calculator.

Practice Exercises

1. What is $\dfrac{2x+6}{x^2-2x-15}$ reduced to simplest terms?

(1) $\dfrac{6}{x^2-15}$

(2) $\dfrac{2x}{x^2-15}$

(3) $\dfrac{2}{x-5}$

(4) $\dfrac{6}{2x-3}$

2. What is $\dfrac{12-3x}{5x-5} \cdot \dfrac{x^2-1}{x^2-3x-4}$ reduced to simplest terms?

(1) $\dfrac{3}{5}$

(2) $\dfrac{12}{5}$

(3) $-\dfrac{3}{5}$

(4) $-\dfrac{12}{5}$

3. What is $\dfrac{4x+16}{x^2+5x+6} \div \dfrac{x^2-16}{5x+15}$ reduced to simplest terms?

(1) $\dfrac{20}{x+2}$

(2) $\dfrac{20}{x-4}$

(3) $\dfrac{20}{(x+2)(x+3)}$

(4) $\dfrac{20}{(x+2)(x-4)}$

4. What is $\dfrac{4}{(x+2)(x+3)} + \dfrac{3}{(x+2)(x-2)}$?

(1) $\dfrac{7x+1}{(x+2)(x-2)(x+3)}$

(2) $\dfrac{7}{(x+2)(x-2)(x+3)}$

(3) $\dfrac{7x+3}{(x+2)(x-2)(x+3)}$

(4) $\dfrac{7}{(x+2)^2(x-2)(x+3)}$

5. What is $\dfrac{x}{x^2+7x+12} - \dfrac{3}{x^2-9}$?

(1) $\dfrac{x^2-6x+12}{(x+3)(x-3)(x+4)}$

(2) $\dfrac{x-3}{7x+21}$

(3) $\dfrac{x^2-6x-12}{(x+3)(x-3)(x+4)}$

(4) $\dfrac{x-3}{7x+3}$

6. Solve for x.

$$\frac{3}{x+2} + \frac{2}{x-3} = \frac{15}{(x+2)(x-3)}$$

(1) 2
(2) 3
(3) 4
(4) 5

7. Solve for x.

$$\frac{1}{x-4} + \frac{3}{x-1} = \frac{7}{x^2-5x+4}$$

(1) 4
(2) 5
(3) 6
(4) 7

8. A group of people contribute equal amounts of money to get a $24 gift for a friend. If two more people contributed and they all paid equal amounts, they would each pay $1 less. How many people were there originally?

9. What are the asymptotes of the graph of $R(x) = \dfrac{3}{x+4}$?

    (1) $x = 4, y = 0$
    (2) $x = -4, y = 0$
    (3) $x = 0, y = 4$
    (4) $x = 0, y = -4$

10. Which could be the equation for this graph?

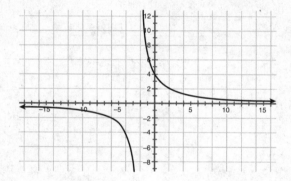

    (1) $R(x) = \dfrac{12}{x+3}$

    (2) $R(x) = \dfrac{x+3}{12}$

    (3) $R(x) = \dfrac{x-3}{12}$

    (4) $R(x) = \dfrac{12}{x-3}$

Solutions

1. Factor the numerator and denominator, and cancel the common factor $(x + 3)$.

$$\frac{2x+6}{x^2-2x-15} = \frac{2(x+3)}{(x+3)(x-5)} = \frac{2}{x-5}$$

The correct choice is **(3)**.

2. Factor both numerators and denominators, and cancel the common factors. A factor from the numerator in one of the fractions can be cancelled with a denominator in the other fraction.

$$\frac{12-3x}{5x-5} \cdot \frac{x^2-1}{x^2-3x-4} = \frac{3(4-x)}{5(x-1)} \cdot \frac{(x-1)(x+1)}{(x-4)(x+1)}$$

Cancel the $(x + 1)$s and the $(x - 1)$s. Then, since $(4 - x) = -1(x - 4)$, the $(4 - x)$ in the numerator of the first fraction can be canceled out with the $(x - 4)$ in the second fraction, leaving a -1 in the numerator.

$$\frac{3(4-x)}{5} \cdot \frac{1}{(x-4)} = \frac{3(-1)}{5} = -\frac{3}{5}$$

The correct choice is **(3)**.

3. Divide by multiplying the first expression by the reciprocal of the second.

$$\frac{4x+16}{x^2+5x+6} \cdot \frac{5x+15}{x^2-16} = \frac{4(x+4)}{(x+2)(x+3)} \cdot \frac{5(x+3)}{(x-4)(x+4)}$$

Cancel the $(x + 3)$s and the $(x + 4)$s.

$$\frac{4}{(x+2)} \cdot \frac{5}{(x-4)} = \frac{20}{(x+2)(x-4)}$$

The correct choice is **(4)**.

4. Since the two denominators have a common factor of $(x + 2)$, this factor is only used once in the common denominator, $(x + 2)(x - 2)(x + 3)$

$$\frac{4(x-2)}{(x+2)(x-2)(x+3)} + \frac{3(x+3)}{(x+2)(x-2)(x+3)}$$

$$= \frac{4(x-2)+3(x+3)}{(x+2)(x-2)(x+3)}$$

$$= \frac{4x-8+3x+9}{(x+2)(x-2)(x+3)}$$

$$= \frac{7x+1}{(x+2)(x-2)(x+3)}$$

The correct choice is **(1)**.

5. First factor the two denominators.

$$\frac{x}{(x+3)(x+4)} - \frac{3}{(x+3)(x-3)}$$

Since the denominators have a common factor of $(x + 3)$, this factor is only used once in the common denominator $(x + 3)(x - 3)(x + 4)$.

$$\frac{x(x-3)}{(x+3)(x-3)(x+4)} - \frac{3(x+4)}{(x+3)(x-3)(x+4)}$$

$$= \frac{x(x-3)-3(x+4)}{(x+3)(x-3)(x+4)}$$

$$= \frac{x^2 - 3x - 3x - 12}{(x+3)(x-3)(x+4)}$$

$$= \frac{x^2 - 6x - 12}{(x+3)(x-3)(x+4)}$$

The correct choice is **(3)**.

6. The common denominator for all three terms is $(x + 2)(x - 3)$. Convert all terms to the common denominator, combine the two terms on the left side of the equal sign, and create an equation to find the value of x that makes the two numerators equal. Check your answer by substituting it back into the original equation.

$$\frac{3(x-3)}{(x+2)(x-3)} + \frac{2(x+2)}{(x+2)(x-3)} = \frac{15}{(x+2)(x-3)}$$

$$\frac{3(x-3)+2(x+2)}{(x+2)(x-3)} = \frac{15}{(x+2)(x-3)}$$

$$\frac{3x-9+2x+4}{(x+2)(x-3)} = \frac{15}{(x+2)(x-3)}$$

$$3x-9+2x+4 = 15$$

$$5x-5 = 15$$

$$5x = 20$$

$$x = 4$$

The correct choice is **(3)**.

7. Factor the denominator of the fraction on the right side of the equal sign into $(x - 4)(x - 1)$. The common denominator for all three terms is $(x - 4)(x - 1)$. Convert all terms to the common denominator, combine the two terms on the left side of the equal sign, and create an equation to find the value of x that makes the two numerators equal. Check your answer by substituting it back into the original equation.

$$\frac{1(x-1)}{(x-4)(x-1)} + \frac{3(x-4)}{(x-4)(x-1)} = \frac{7}{(x-4)(x-1)}$$

$$\frac{1(x-1)+3(x-4)}{(x-4)(x-1)} = \frac{7}{(x-4)(x-1)}$$

$$\frac{x-1+3x-12}{(x-4)(x-1)} = \frac{7}{(x-4)(x-1)}$$

$$x-1+3x-12 = 7$$

$$4x-13 = 7$$

$$4x = 20$$

$$x = 5$$

The correct choice is **(2)**.

8. 6 people

If there were originally x people, the amount each person would pay is $\frac{24}{x}$. If two more people were involved, there would be $x + 2$ people so the amount each person would pay is $\frac{24}{x+2}$. Since each person's share would be $1 less if there were two more people, the equation relating these expressions is $\frac{24}{x} = \frac{24}{x+2} + 1$. This can be solved by making all three terms have a common denominator of $x(x + 2)$ and solving.

$$\frac{24(x+2)}{x(x+2)} = \frac{24x}{x(x+2)} + \frac{x(x+2)}{x(x+2)}$$

$$\frac{24x+48}{x(x+2)} = \frac{24x+x^2+2x}{x(x+2)}$$

$$24x+48 = 24x+x^2+2x$$

$$0 = x^2+2x-48$$

$$0 = (x+8)(x-6)$$

$$x+8 = 0 \quad \text{or} \quad x-6 = 0$$

$$x = -8 \quad \text{or} \quad x = 6$$

Since the amount each pays must be positive, the -8 is rejected, and the only answer is $x = 6$.

9. One way is to graph the function on the graphing calculator and see the curve approach the x-axis, which has equation $y = 0$, for large x-values and approach the vertical line $x = -4$ for values close to -4.

Even without the graphing calculator, you can find the equation of the vertical asymptote(s) by setting the denominator equal to zero.

$$x + 4 = 0$$
$$x = -4$$

This is the equation of the vertical asymptote and only choice (2) has this as one of the asymptotes.

The correct choice is **(2)**.

10. Since there is a vertical asymptote at $x = 3$ on the graph, there must be a $x - 3$ in the denominator of the rational expression. This only happens in choice (4).

The correct choice is **(4)**.

3. EXPONENTIAL AND LOGARITHMIC EXPRESSIONS AND EQUATIONS

3.1 PROPERTIES OF EXPONENTS

In an exponential expression like $3^5 = 3 \cdot 3 \cdot 3 \cdot 3 \cdot 3 = 243$ the 3 is called the *base* and the 5 is called the **exponent** or the *power*.

To multiply two exponential expressions with the same base like $3^8 \cdot 3^2$, the product has the same base as the factors do and the power of the product will be the sum (and *not* the product) of the powers of the factors.

$$3^8 \cdot 3^2 = 3^{8+2} = 3^{10}$$

USEFUL FORMULA

$$x^a \cdot x^b = x^{a+b}$$

- Division is similar except you subtract the two exponents instead of adding them.

$$3^8 \div 3^2 = 3^{8-2} = 3^6$$

USEFUL FORMULA

$$x^a \div x^b = x^{a-b}$$

$$\frac{x^a}{x^b} = x^{a-b}$$

- To raise an exponential expression to a power, like $(3^8)^2$, multiply the two exponents together.

$$(3^8)^2 = 3^{8 \times 2} = 3^{16}$$

USEFUL FORMULA

$$(x^a)^b = x^{ab}$$

- To raise a number or expression to a negative exponent, make a fraction with a numerator of 1 and a denominator with the same expression, but with the negative sign changed to a positive sign.

$$5^{-3} = \frac{1}{5^{+3}} = \frac{1}{125}$$

USEFUL FORMULA

$$x^{-a} = \frac{1}{x^a}$$

- To raise a number or expression to a fractional exponent, like $8^{\frac{5}{3}}$, take the third root of 8 and raise the answer to the fifth power.

$$8^{\frac{5}{3}} = \left(\sqrt[3]{8}\right)^5 = 2^5 = 32$$

USEFUL FORMULA

$$x^{\frac{a}{b}} = \left(\sqrt[b]{x}\right)^a$$

Anything but 0 raised to the 0th power is 1, for example $8^0 = 1$.

It is *not* true that $(5 + 3)^2 = 5^2 + 3^2$ since the left side is equal to 64 while the right side is equal to 34.

There is, though, a distributive property of exponents when a single term with multiple factors is raised to a power like $(5 \cdot 3)^2 = 5^2 \cdot 3^2$.

3.2 EXPONENTIAL EQUATIONS

If an equation involves two exponential equations with the same base, the exponents must be equal.

To solve for x in the equation $4^{2x + 1} = 4^{3x - 5}$, solve the equation $2x + 1 = 3x - 5$ so $x = 6$.

If the bases are different, it is sometimes possible to convert one or both of the expressions so they do have the same base.

Example

Solve for x in $8^{x + 4} = 2^{5x - 4}$.

Solution:

Notice that 8 can be replaced with 2^3.

$$8^{x + 4} = 2^{5x - 4}$$
$$2^{3(x + 4)} = 2^{5x - 4}$$
$$3(x + 4) = 5x - 4$$
$$3x + 12 = 5x - 4$$
$$16 = 2x$$
$$8 = x$$

3.3 LOGARITHMS

A logarithmic expression looks like $\log_2 64$. This can be evaluated to a number by thinking "To what power must the number 2 be raised in order to get a value of 64?" Since $2^6 = 64$, $\log_2 64 = 6$.

USEFUL FORMULA

A logarithmic equation like $\log_a b = c$ can be rearranged into an exponential equation $a^c = b$.

Example

Write the equation $\log_7 2{,}401 = x$ as an exponential equation. Then solve that equation by guess-and-check.

Solution:

The equation is $7^x = 2{,}401$. By testing different positive integers, it is quickly found that $7^4 = 2{,}401$ so the solution is $x = 4$.

The TI-84 and the TI-Nspire calculators have logarithm (or, for short, log) functions, which can be used to solve logarithmic equations.

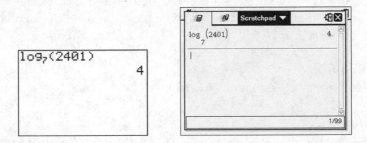

Example

Convert the equation $5^x = 700$ into a logarithmic equation, and then use the graphing calculator to approximate the answer to the nearest hundredth.

Solution:

The equation is $\log_5 700 = x$. Using the log function of your calculator, $x = 4.07$.

If, in an exponential equation, there is a number multiplied by the exponential expression, divide both sides of the equation by that number before converting the equation to a log equation.

Example
Solve for x in the equation $800 = 200 \cdot 1.04^x$.

Solution:
Do not multiply 200 and 1.04. Instead first divide both sides of the equation by 200 to get $4 = 1.04^x$. This can then be solved with the graphing calculator as $\log_{1.04} 4 = 35.35$.

- The number e is a mathematical constant that is equal to approximately 2.72. Many exponential equations have a base of e. Since e is such a popular base, there is a special key on the calculator, the LN key, which can be used to calculate log base e.

Example
Solve for x in the equation $e^x = 150$.

Solution:
First rewrite as $\log_e 150 = x$. This can be solved with the graphing calculator by either doing $\log_e 150 = 5.01$ or, even shorter, $\ln 150 = 5.01$.

3.4 GRAPHS OF EXPONENTIAL AND LOGARITHMIC FUNCTIONS

The graph of $y = 2^x$ looks like a playground slide going up to the right. When the base of the exponent is less than 1, like in $y = \left(\frac{1}{2}\right)^x$, it looks like a playground slide going down to the right.

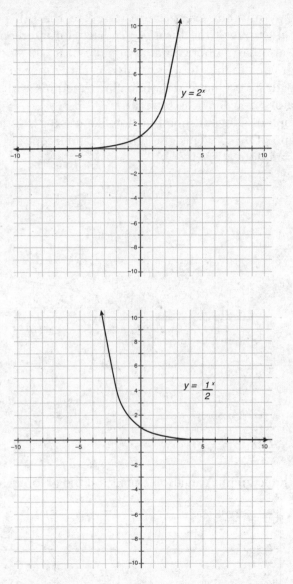

Since any number except zero raised to the 0th power is 1, exponential graphs of the form $y = b^x$ have a y-intercept of $(0, 1)$. They will also have a horizontal asymptote of $y = 0$ since raising a number greater than 1 to a very large negative power becomes a fraction with a very large denominator. Fractions with very large denominators are very close to zero.

The graph of a logarithmic function like $y = \log_2 x$ is a reflection of the graph of $y = 2^x$. This graph has an x-intercept of $(1, 0)$ and a vertical asymptote at $x = 0$.

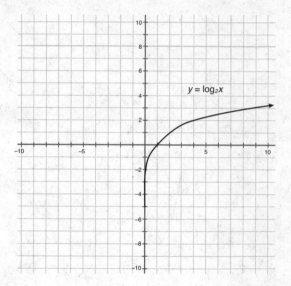

$y = \log_2 x$

Logarithmic graphs can also be made on the graphing calculator.

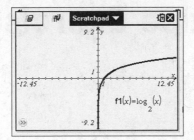

3.5 MODELING REAL-WORLD SCENARIOS WITH EXPONENTIAL EQUATIONS

Many real-world problems involve things that grow or decay in a way that can be modeled with exponential equations.

Example

The population of a country can be modeled by the equation $P = 50(1.02)^T$, where T is the number of years since 2010 and P is the population in millions. In what year will the population reach 75 million?

Solution:

$$75 = 50(1.02)^T$$
$$\frac{75}{50} = \frac{50(1.02)^T}{50}$$
$$1.5 = 1.02^T$$
$$T = \log_{1.02} 1.5 = 20.5$$

It will reach the population of 75 million 20.5 years after 2010, or sometime between 2030 and 2031.

Practice Exercises

1. What is $a^{\frac{3}{4}}$?

   (1) $\sqrt[3]{a^4}$

   (2) $a^{-\frac{3}{4}}$

   (3) $a^{-\frac{4}{3}}$

   (4) $\left(\sqrt[4]{a^{-3}}\right)$

2. Simplify $(2x^3y^4)^3$.

   (1) $8x^{27}y^{64}$

   (2) $6x^9y^{12}$

   (3) $8x^9y^{12}$

   (4) $6x^{27}y^{64}$

3. Solve $4^x = 256$.

   (1) 4

   (2) 8

   (3) 16

   (4) 64

4. Solve $8^{x+4} = 2^{x+4}$.

   (1) 2

   (2) –2

   (3) 1

   (4) –4

5. The equation $m^x = n$ is equivalent to which of the following?

   (1) $\log_n m = x$

   (2) $\log_m x = n$

   (3) $\log_m n = x$

   (4) $\log_n x = m$

6. What is the solution to x to the *nearest* tenth in $3 \cdot 4^x = 270$?
   (1) 3.0
   (3) 3.2
   (2) 3.1
   (4) 3.3

7. If $x = \log_7 7^{123456789}$, what is the value of x?

8. The population of West Algebra can be modeled by the equation $P = 30 \cdot 1.04^T$ where T is the number of years since 2000 and P is the population in millions. How many million people will there be in 2020?
   (1) 63.7
   (3) 65.7
   (2) 64.7
   (4) 66.7

9. The temperature of a cup of herbal tea can be modeled by the equation $T = 90 \cdot 0.7^M + 75$ where T is the temperature and M is the number of minutes since the tea was taken off the stove. How hot will the tea be 15 minutes after it is taken off the stove?
   (1) 75°
   (3) 79°
   (2) 77°
   (4) 81°

10. Which equation is equivalent to the equation $y = 50(1.07)^{12x}$?
    (1) $y = 50(2.1)^x$
    (3) $y = 50(1.3)^{6x}$
    (2) $y = 50(1.5)^{2x}$
    (4) $y = 53.5^{12x}$

Solutions

1. The denominator of the fraction goes outside the radical sign and the numerator becomes the exponent that the radical gets raised to.

$$a^{\frac{3}{4}} = (\sqrt[4]{a})^3$$

The correct choice is **(4)**.

2. Each factor in the parentheses gets raised to the 3rd power. When something raised to a power is then raised to another power, the base does not change but the exponent becomes the product of the two exponents.

$$(2x^3y^4)^3 = 2^3(x^3)^3(y^4)^3 = 8x^9y^{12}$$

The correct choice is **(3)**.

3. Because this is a multiple-choice question, you can test the four choices and find that for choice (1), $4^4 = 256$.

The correct choice is **(1)**.

4. If possible, change both expressions so they have the same base. Since $8 = 2^3$, this equation can be written as

$$(2^3)^{x + 4} = 2^{x + 4}$$
$$2^{3(x + 4)} = 2^{x + 4}$$

If the bases are equal and the expressions are equal, the exponents must be equal. This lead to a new equation.

$$3(x + 4) = x + 4$$
$$3x + 12 = x + 4$$
$$2x = -8$$
$$x = -4$$

The correct choice is **(4)**.

5. If $m^x = n$, then by the definition of logs, $\log_m n = x$.

The correct choice is **(3)**.

6. First eliminate the 3 by dividing both sides by 3.

$$\frac{3 \cdot 4^x}{3} = \frac{270}{3}$$
$$4^x = 90$$

Rearrange this exponential equation into a log equation, and solve with the log function of the graphing calculator. For the TI-84, press [MATH] [A]. For the TI-Nspire, press [log].

$$\log_4 90 = x$$

$x = 3.2459$, which, rounded to the nearest tenth, is 3.2.

The correct choice is **(3)**.

7. 123456789

This can be rearranged to the exponential equation

$$7^x = 7^{123456789}$$
$$x = 123456789$$

8. Substitute $T = 20$ into the equation, and calculate with your calculator.

$$P = 30 \cdot 1.04^{20}$$
$$P = 65.7$$

The correct choice is **(3)**.

9. Substitute $M = 15$ into the equation, and calculate with your calculator.

$$T = 90 \cdot 0.7^{15} + 75$$
$$T = 75°$$

The correct choice is **(1)**.

10. The rule for raising a power to a power is $(x^a)^b = x^{ab}$. This rule also works in reverse $x^{ab} = (x^a)^b$. So $(1.07)^{12x} = (1.07^{12})^x$. Calculate $1.07^{12} = 2.25$ so one equivalent equation is $y = 50(2.25)^x$. Unfortunately, this is not one of the choices. But if you do this process for each of the choices you will see that choice (2), $y = 50(1.5)^{2x} = 50(1.5^2)^x = 50(2.25)^x$ is the correct answer.

The correct choice is **(2)**.

4. RADICAL EXPRESSIONS AND EQUATIONS

4.1 SIMPLIFYING RADICALS

The square root of a perfect square, like $\sqrt{25}$, will be a positive integer. In this case, $\sqrt{25} = 25$ because $5^2 = 25$.

- The square root of something that is not a perfect square, like $\sqrt{27}$, will be an irrational number. In this case $\sqrt{27}$ will be between 5 and 6 because $5^2 = 25$ and $6^2 = 36$ and 27 is between 25 and 36. Using a calculator you can find that $\sqrt{27} \approx 5.196$.

- Radical expressions like $\sqrt{7} \cdot \sqrt{11}$ can be multiplied by multiplying the numbers inside the radicals together, $\sqrt{7} \cdot \sqrt{11} = \sqrt{7 \cdot 11} = \sqrt{77}$. This process can also be reversed when simplifying the square root of any composite number like $\sqrt{8} = \sqrt{2 \cdot 4} = \sqrt{2} \cdot \sqrt{4}$.

If the number inside the square root sign has a factor that is a perfect square, the expression can be simplified. Since $27 = 9 \cdot 3$ and 9 is a perfect square, $\sqrt{27} = \sqrt{9 \cdot 3} = \sqrt{9} \cdot \sqrt{3} = 3 \cdot \sqrt{3} = 3\sqrt{3}$.

Two radical expressions can be added or subtracted if the number inside the radical is the same, so $7\sqrt{2} + 3\sqrt{2} = 10\sqrt{2}$ and $7\sqrt{2} - 3\sqrt{2} = 4\sqrt{2}$.

4.2 IMAGINARY AND COMPLEX NUMBERS

The number i is defined as $\sqrt{-1}$.

USEFUL FACTS

The first four powers of i are:

$$i^1 = i, i^2 = -1, i^3 = i^2 \cdot i^1 = -1 \cdot i = -i,$$
$$\text{and } i^4 = i^2 \cdot i^2 = (-1) \cdot (-1) = 1$$

Anytime a negative number is inside a square root sign it can be rewritten with an i. Because of this definition, $i^2 = \left(\sqrt{-1}\right)^2 = -1$.

$$\sqrt{-9} = \sqrt{9} \cdot \sqrt{-1} = 3i$$

A number of the form ai, where a is a real number, is known as an **imaginary number**.

- Imaginary numbers can be added or subtracted

$$5i + 2i = 7i$$
$$5i - 2i = 3i$$

- Imaginary numbers can be multiplied. Whenever an i^2 appears in a solution, replace it with a -1.

$$5i \cdot 2i = 10i^2 = 10 \cdot (-1) = -10$$

A number of the form $a + bi$, where a and b are real numbers, is called a **complex number**. $5 + 3i$ is a complex number.

Complex numbers can be added and subtracted

$$(5 + 3i) + (2 + 7i) = 7 + 10i$$
$$(5 + 3i) - (2 + 7i) = 3 - 4i$$

- Complex numbers can be multiplied. Whenever an i^2 appears in a solution, replace it with a -1.

$$(5 + 3i)(2 + 7i) = 10 + 35i + 6i + 21i^2 = 10 + 41i - 21 = -11 + 41i$$

Sometimes complex numbers appear when solving quadratic equations with the quadratic formula.

Example
Solve for all solutions in the equation $x^2 - 4x + 13 = 0$.

Solution:
Using the quadratic equation, $a = 1$, $b = -4$, and $c = 13$.

$$x = \frac{-(-4) \pm \sqrt{(-4)^2 - 4(1)(13)}}{2(1)} = \frac{4 \pm \sqrt{16 - 52}}{2} = \frac{4 \pm \sqrt{-36}}{2}$$

Since there is a negative number inside the radical sign, the solutions will be complex numbers.

$$x = \frac{4 \pm 6i}{2} = 2 \pm 3i$$

There are two complex solutions $x = 2 + 3i$ or $x = 2 - 3i$.

- Complex numbers can be graphed as points on **the complex plane**.

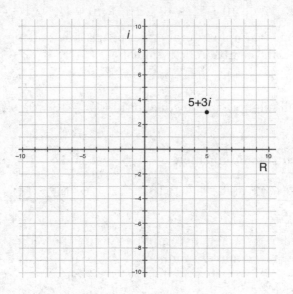

USEFUL FORMULA

The absolute value of a complex number $a + bi$ is the distance of the point that represents that number on the complex plane from the point that represents the number $0 + 0i$. The formula is $|a + bi| = \sqrt{a^2 + b^2}$.

The absolute value of $5 + 3i$ is $\sqrt{5^2 + 3^2} = \sqrt{25 + 9} = \sqrt{34}$.

4.3 RADICAL EQUATIONS

If an equation has one radical sign, isolate it and then square both sides of the equation to get a new equation. The solutions to that new equation may be solutions to the original equation.

Example

Solve for all values of x that satisfy the equation $\sqrt{x} - x = -6$.

Solution:

- **Step One:** Isolate the radical.

$$\sqrt{x} - x = -6$$
$$\underline{+x = +x}$$
$$\sqrt{x} = x - 6$$

- **Step Two:** Square both sides.

$$\left(\sqrt{x}\right)^2 = (x - 6)^2$$
$$x = x^2 - 12x + 36$$
$$0 = x^2 - 13x + 36$$
$$0 = (x - 4)(x - 9)$$

$$x - 4 = 0 \quad \text{or} \quad x - 9 = 0$$
$$x = 4 \quad \text{or} \quad x = 9$$

- **Step Three:** Substitute answers back into the original equation to check if some solutions should be rejected.

$$\sqrt{4} - 4 = 2 - 4 = -2$$
$$\sqrt{9} - 9 = 3 - 9 = -6$$

The 4 must be rejected since it does not satisfy the original equation so the only solution is $x = 9$.

Solving an equation that involves two radical terms like $\sqrt{x+1} + \sqrt{x-4} = 5$ is a lengthy process. To solve it, first isolate one of the radical terms, square both sides, and continue with the problem by isolating the other radical term.

$$\sqrt{x+1} + \sqrt{x-4} = 5$$
$$\underline{-\sqrt{x-4} = -\sqrt{x-4}}$$
$$\sqrt{x-4} = 5 - \sqrt{x-4}$$
$$\left(\sqrt{x+1}\right)^2 = \left(5 - \sqrt{x-4}\right)^2$$
$$x+1 = 25 - 10\sqrt{x-4} + x - 4$$
$$\underline{-x = -x}$$
$$1 = 21 - 10\sqrt{x-4}$$
$$\underline{-21 = -21}$$
$$-20 = -10\sqrt{x-4}$$
$$\frac{-20}{-10} = \frac{-10\sqrt{x-4}}{-10}$$
$$2 = \sqrt{x-4}$$
$$2^2 = \left(\sqrt{x-4}\right)^2$$
$$4 = x - 4$$
$$\underline{+4 = +4}$$
$$8 = x$$

Check to see if the answer should be rejected.

$$\sqrt{8+1} + \sqrt{8-4} = \sqrt{9} + \sqrt{4} = 3 + 2 = 5$$

The solution is $x = 8$.

4.4 GRAPHS OF RADICAL FUNCTIONS

The graph of $y = \sqrt{x}$ is half of a sideways parabola.

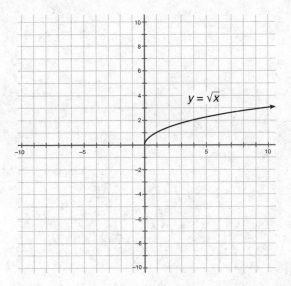

Practice Exercises

1. What is $\sqrt{12} + \sqrt{75}$?

   (1) $7\sqrt{3}$ (3) $7\sqrt{6}$

   (2) $\sqrt{87}$ (4) $3\sqrt{7}$

2. What is $2\sqrt{3} \cdot 5\sqrt{3}$?

   (1) 30   (3) $\sqrt{30}$

   (2) $10\sqrt{3}$ (4) $7\sqrt{3}$

3. What is i^{66}?
   (1) -1 (3) $-i$
   (2) 1   (4) i

4. What is $(3 + 5i) + (2 - 7i)$?
   (1) $5 + 2i$ (3) $-5 - 2i$
   (2) $-5 + 2i$ (4) $5 - 2i$

5. Solve the quadratic equation $x^2 - 6x + 34 = 0$.
   (1) $3 \pm 5i$ (3) $13, -7$
   (2) $-3 \pm 5i$ (4) $-13, 7$

6. What is $|12 - 5i|$?

    (1) $\sqrt{119}$ (3) 13

    (2) 12   (4) 7

7. What value(s) of x make the equation $\sqrt{x} + x = 20$ true?
    (1) 14   (3) 16
    (2) 25   (4) 16, 25

8. Find all solutions to the equation $\sqrt{x+3} + x = 9$.
    (1) 13, 6   (3) 6
    (2) 13   (4) 4

9. Solve for x.

$$\sqrt{x-3} + \sqrt{x+9} = 6$$

    (1) 6   (3) 8
    (2) 7   (4) 9

10. Solve for x.

$$5\sqrt{x-4} + 3 = 18$$

Solutions

1. Simplify each radical and, if they have the same number under the radical after simplifying, combine them.

$$\sqrt{12} + \sqrt{75} = \sqrt{4 \cdot 3} + \sqrt{25 \cdot 3}$$
$$= \sqrt{4}\sqrt{3} + \sqrt{25}\sqrt{3}$$
$$= 2\sqrt{3} + 5\sqrt{3}$$
$$= 7\sqrt{3}$$

The correct choice is (**1**).

2. $2\sqrt{3} \cdot 5\sqrt{3} = 2 \cdot 5 \cdot \sqrt{3} \cdot \sqrt{3} = 10 \cdot 3 = 30$

The correct choice is (**1**).

3. Since $i^4 = 1$, $i^{64} = (i^4)^{16} = 1^{16} = 1$. So i raised to any multiple of 4 will be 1. 66 is not a multiple of 4, but it can be written as $64 + 2$ so $i^{66} = i^{64 + 2} = i^{64} \cdot i^2 = 1 \cdot i^2 = -1$.

The correct choice is (**1**).

4. Combine like terms

$(3 + 5i) + (2 - 7i) = 3 + 5i + 2 - 7i = 3 + 2 + 5i - 7i = 5 - 2i$

The correct choice is (**4**).

5. Using the quadratic formula with $a = 1$, $b = -6$, and $c = 34$,

$$x = \frac{-(-6) \pm \sqrt{(-6)^2 - 4(1)(34)}}{2(1)}$$

$$= \frac{6 \pm \sqrt{36 - 136}}{2}$$

$$= \frac{6 \pm \sqrt{-100}}{2}$$

$$= \frac{6 \pm 10i}{2}$$

$$= 3 \pm 5i$$

The correct choice is **(1)**.

6. The absolute value of the complex number $a + bi$ is

$$|a + bi| = \sqrt{a^2 + b^2}$$

So $|12 - 5i| = \sqrt{12^2 + (-5)^2} = \sqrt{144 + 25} = \sqrt{169} = 13$.

The correct choice is **(3)**.

7. Isolate the radial expression, and square both sides.

$$\sqrt{x} + x = 20$$
$$\underline{-x = -x}$$
$$\sqrt{x} = -x + 20$$
$$\left(\sqrt{x}\right)^2 = (-x + 20)^2$$
$$x = x^2 - 40x + 400$$
$$\underline{-x = -x}$$
$$0 = x^2 - 41x + 400$$
$$0 = (x - 16)(x - 25)$$

$$x - 16 = 0 \quad \text{or} \quad x - 25 = 0$$
$$x = 16 \quad \text{or} \quad x = 25$$

Test each of these in the original equation to make sure that an extraneous solution was not created when both sides were squared.

For $x = 16$, $\sqrt{16} + 16 = 4 + 16 = 20$.

For $x = 25$, $\sqrt{20} + 25 = 5 + 25 = 30 \neq 20$.

The only solution is $x = 16$

The correct choice is **(3)**.

8. Isolate the radical expression, square both sides, and solve the resulting equation.

$$\sqrt{x+3} + x = 9$$
$$\underline{-x = -x}$$
$$\sqrt{x+3} = -x + 9$$
$$\left(\sqrt{x+3}\right)^2 = \left(-x+9\right)^2$$
$$x + 3 = x^2 - 18x + 81$$
$$0 = x^2 - 19x + 78$$
$$0 = (x - 13)(x - 6)$$

$$x - 13 = 0 \quad \text{or} \quad x - 6 = 0$$
$$x = 13 \quad \text{or} \quad x = 6$$

Check both of these answers to see if they satisfy the original equation.

For $x = 13$, $\sqrt{13+3} + 13 = \sqrt{16} + 13 = 4 + 13 = 17 \neq 9$.

For $x = 6$, $\sqrt{6+3} + 6 = \sqrt{9} + 6 = 3 + 6 = 9$.

The only solution is $x = 6$.

The correct choice is **(3)**.

9. Isolate one radical; square both sides. Then isolate the other radical and square both sides. Solve the resulting equation, and check your answers for extraneous roots.

$$\sqrt{x-3} + \sqrt{x+9} = 6$$
$$\underline{-\sqrt{x+9} = -\sqrt{x+9}}$$
$$\sqrt{x-3} = 6 - \sqrt{x+9}$$
$$\left(\sqrt{x-3}\right)^2 = \left(6 - \sqrt{x+9}\right)^2$$
$$x - 3 = 36 - 12\sqrt{x+9} + x + 9$$
$$x - 3 = 45 - 12\sqrt{x+9} + x$$
$$\underline{-45 - x = -45 - x}$$
$$-48 = -12\sqrt{x+9}$$
$$\frac{-48}{-12} = \frac{-12\sqrt{x+9}}{-12}$$
$$4 = \sqrt{x+9}$$
$$4^2 = \left(\sqrt{x+9}\right)^2$$
$$16 = x + 9$$
$$\underline{-9 = -9}$$
$$7 = x$$

Check $x = 7$ in the original equation,

$$\sqrt{7-3} + \sqrt{7+9} = \sqrt{4} + \sqrt{16} = 2 + 4 = 6$$

The correct choice is **(2)**.

10.
$$5\sqrt{(x-4)} + 3 = 18$$
$$\underline{-3 = -3}$$
$$\frac{5\sqrt{(x-4)}}{5} = \frac{15}{5}$$
$$\sqrt{(x-4)} = 3$$
$$\left(\sqrt{(x-4)}\right)^2 = 3^2$$
$$x - 4 = 9$$
$$\underline{+4 = +4}$$
$$x = 13$$

5. TRIGONOMETRIC EXPRESSIONS AND EQUATIONS

5.1 UNIT CIRCLE TRIGONOMETRY

A **unit circle** is a circle centered at $(0, 0)$ that has a radius of one unit.

If P is a point on the unit circle and point A is at $(1, 0)$, then $\angle AOP$ is said to be in **standard position**. In the diagram below $\mathrm{m}\angle AOP = 27°$.

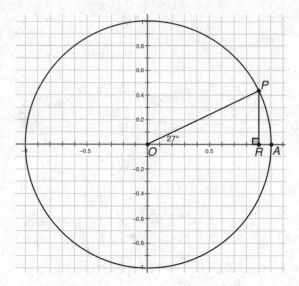

Because triangle OPR is a right triangle, the x-coordinate and y-coordinate of point P can be found with trigonometry. The x-coordinate will always equal $\cos(\angle AOP)$ and the y-coordinate will always equal $\sin(\angle AOP)$. In the following diagram, the coordinates of P are $(\cos 27°, \sin 27°) = (0.89, 0.45)$

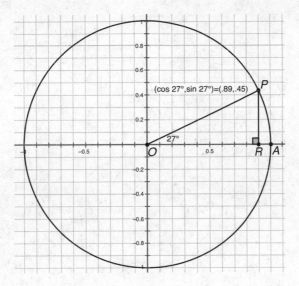

When point P is in quadrant II, the x-coordinate is negative so $\cos(\angle AOP) < 0$.

When point P is in quadrant III, both the x-coordinate and the y-coordinate are negative so $\cos(\angle AOP) < 0$ and $\sin(\angle AOP) < 0$.

When point P is in quadrant IV, the y-coordinate is negative so $\sin(\angle AOP)$ is negative.

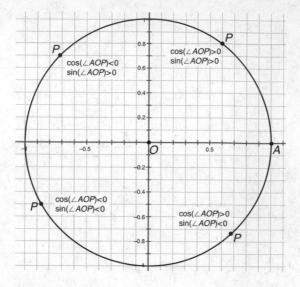

- If the x-coordinate of a point P on the unit circle is known, the $\sin^{-1}$ function on the calculator can be used as a step in finding the measure of $\angle AOP$. If the y-coordinate of a point P on the unit circle is known, the $\cos^{-1}$ function on the calculator can be used as a step in finding the measure of $\angle AOP$.

Example
If the coordinates of P on the unit circle are (0.48, 0.87), what is $m\angle AOP$?

Solution:
P is in quadrant I so $m\angle AOP$ is between $0°$ and $90°$. Since the x-coordinate of the point is 0.48, $\angle AOP = \cos^{-1}(0.48) = 61°$. You could also use the y-coordinate to find $\angle AOP = \sin^{-1}(0.48) = 39°$.

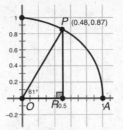

- If point P is not in quadrant I, you find the measure of the *reference angle* by taking either the $\sin^{-1}(|y\text{-coordinate}|)$ or $\cos^{-1}(|x\text{-coordinate}|)$. $\angle AOP$ can be calculated by following one of the four rules.

Quadrant of P	$\angle AOP$
I	reference angle
II	180° − reference angle
III	180° + reference angle
IV	360° − reference angle

Example

If $\sin(\angle AOP) = -\dfrac{4}{5}$ and $\cos(\angle AOP) < 0$, what is m$\angle AOP$?

Solution:

Since the x-coordinate is the cosine, point P has a negative x-coordinate. Since the y-coordinate is the sine, point P has a negative y-coordinate. This means that point P is in quadrant III.

The reference angle is $\sin^{-1}\left(\left|-\dfrac{4}{5}\right|\right) = \sin^{-1}\left(\dfrac{4}{5}\right) = 53.1°$. In quadrant III, $\angle AOP = 180° +$ reference angle $= 180° + 53.1° = 223.1°$.

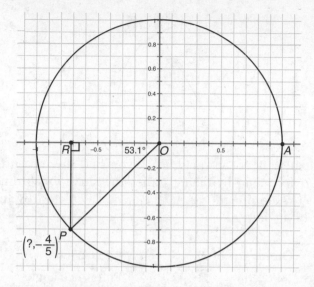

5.2 RADIAN MEASURE

Just as there are different units for measuring lengths, like inches and meters, there are different units for measuring angles. The most common unit for measuring angles is the degree. There are 360 degrees in a circle.

A **radian** is a unit of measurement that is much larger than a degree. A radian is approximately 57.3 degrees.

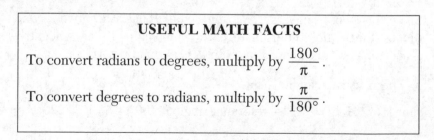

USEFUL MATH FACTS

To convert radians to degrees, multiply by $\dfrac{180°}{\pi}$.

To convert degrees to radians, multiply by $\dfrac{\pi}{180°}$.

Example

Convert $\frac{\pi}{3}$ radians to degrees.

Solution:

$\frac{\pi}{3}$ radians is equal to $\frac{\pi}{3} \cdot \frac{180°}{\pi} = 60°$.

USEFUL MATH FACTS

Some common radian to degree conversions are:

Degrees	Radians
360°	2π
180°	π
90°	$\frac{\pi}{2}$
60°	$\frac{\pi}{3}$
45°	$\frac{\pi}{4}$
30°	$\frac{\pi}{6}$

5.3 GRAPHS OF THE SINE AND COSINE FUNCTIONS

The graph of $y = \sin x$ looks like this.

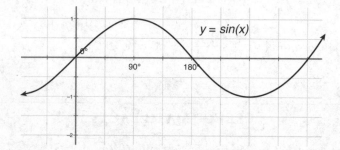

The graph of $y = \cos x$ looks like this

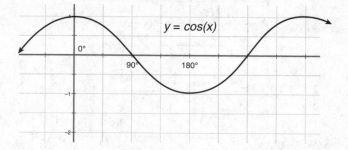

Curves that look like these are known as **sinusoidal curves** and have a **midline**, an **amplitude**, and a **period**. The midline is the invisible horizontal line passing through the middle of the curve. The amplitude is the distance between the midline and either a high point or a low point on the curve. The period is the horizontal distance between two consecutive high points or low points.

For $y = \sin x$ and $y = \cos x$, the midlines are $y = 0$, the amplitudes are 1, and the periods are 360° (or 2π radians).

MATH FACTS

The graph of an equation of the form $y = A\sin(Bx) + D$ or $y = A\cos(Bx) + D$ has a midline of $y = D$, an amplitude of A, and a period of $\dfrac{360°}{B}$ (or $\dfrac{2\pi}{B}$ radians). If A is negative, the curve is "upside down."

Example

What is the midline, amplitude, and period of the curve graphed below with equation $y = 3\sin(2x) - 1$?

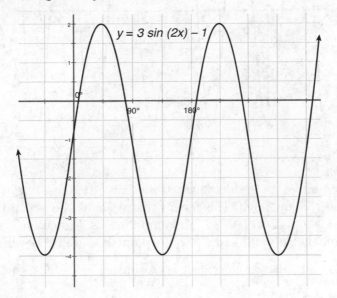

Solution:

The midline is $y = -1$, the amplitude is 3, and the period is $\dfrac{360°}{2} = 180°$.

Example

Below is one cycle of a sinusoidal curve. What could be the equation of this curve?

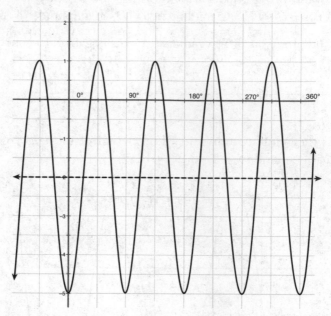

Solution:

This resembles an upside down cosine curve because the "starting point" is the low point. This means that A should be negative. The midline is $y = -2$ so $D = -2$. The amplitude is 3 since the distance between the midline and the high point is 3 units. Because the period is $90°$, the B value must be 4 since $\dfrac{360°}{4} = 90°$. The equation is $y = -3\cos(4x) - 2$.

5.4 TRIGONOMETRY EQUATIONS

An equation that involves one of the trigonometric functions, like sine or cosine, is called a trigonometric equation. Trigonometric equations often have more than one solution. The simplest way to find the solutions to a trigonometric equation is to use a graphing calculator.

Example

Use your graphing calculator to find all solutions between $0° \leq x < 360°$ that satisfy the equation $\sin x = -0.94$.

Solution:

Graph $y = \sin x$ and $y = -0.94$ on the same set of axes. Use the Zoom Trig command to set a useful window. Then use the intersect feature of the calculator to find the x-coordinates of the intersection points that are between 0 and 360.

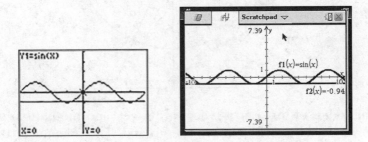

The two intersection points are $(250, -0.94)$ and $(290, -0.94)$ so the solutions are $x = 250$ and $x = 290$.

5.5 MODELING REAL-WORLD SCENARIOS WITH TRIG FUNCTIONS

Certain real-world scenarios can be modeled with a trigonometric equation. The graphing calculator can be used to solve this equation.

Example

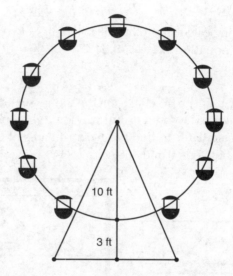

The radius of this Ferris wheel is 10 feet, and the bottom of the Ferris wheel is 3 feet above the ground. The amount of time it takes to make a complete revolution around the Ferris wheel is 100 seconds.

When Evelyn gets on the Ferris wheel, she is exactly 3 feet above the ground. 50 seconds after starting, she will be 23 feet above the ground, at the peak of the Ferris wheel. 100 seconds after starting Evelyn will be back to her starting position, again exactly 3 feet above the ground.

The graph of Evelyn's height above the ground will look like this.

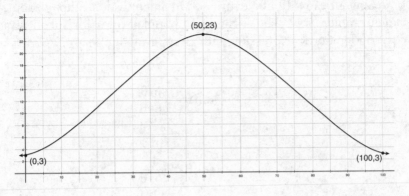

After how many seconds will she be 18 feet above the ground?

Solution:
When the unit on the *x*-axis is not degrees, as in this case where it is seconds, assume everything is in radians.

The equation is $h = -10 \cos\left(\dfrac{\pi}{50}t\right) + 13$. The *B* value was calcu-

lated by using the formula period $= \dfrac{2\pi}{B}$. The period is given as 100

so $100 = \dfrac{2\pi}{B}$, $100B = 2\pi$, $B = \dfrac{2\pi}{100} = \dfrac{\pi}{50}$.

Graphing $y = -10 \cos\left(\dfrac{\pi}{50}x\right) + 13$ and $y = 18$ on the graphing

calculator and finding the *x*-coordinates of the intersection points gets solutions of $x = 33$ and $x = 67$ seconds.

5.6 TRIGONOMETRIC IDENTITIES

An expression involving sine, cosine, tangent, cosecant, secant, or cotangent can often be expressed in an equivalent, though sometimes simpler, way. The eight most popular rules for substituting one trigonometric expression with an equivalent one are:

$$\sin^2\theta + \cos^2\theta = 1$$

$$\tan\theta = \frac{\sin\theta}{\cos\theta}$$

$$\csc\theta = \frac{1}{\sin\theta}$$

$$\sec\theta = \frac{1}{\cos\theta}$$

$$\cot\theta = \frac{1}{\tan\theta}$$

$$1 + \cot^2\theta = \csc^2\theta$$

$$\tan^2\theta + 1 = \sec^2\theta$$

- These identities can be used to check if more complicated expressions involving trigonometric functions are equivalent.

Example

Show that $\sin x \tan x + \cos x = \sec x$.

Solution:

$$\sin x \tan x + \cos x$$

$$= \sin x \frac{\sin x}{\cos x} + \cos x$$

$$= \frac{\sin^2 x}{\cos x} + \frac{\cos x}{1}$$

$$= \frac{\sin^2 x}{\cos x} + \frac{\cos^2 x}{\cos x}$$

$$= \frac{\sin^2 x + \cos^2 x}{\cos x}$$

$$= \frac{1}{\cos x}$$

$$= \sec x$$

Practice Exercises

1. In unit circle O, the coordinates of point B are $(-0.91, -0.41)$. What is the measure of $\angle AOB$?

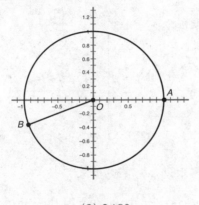

(1) 211°   (3) 246°

(2) 237°   (4) 204°

2. Convert 240° to radians.

(1) $\dfrac{4}{3}$ (3) $\dfrac{4\pi}{3}$

(2) $\dfrac{5\pi}{6}$ (4) $\dfrac{5\pi}{4}$

3. Convert $\frac{\pi}{4}$ radians to degrees.

(1) 90°   (3) 30°

(2) 60°   (4) 45°

4. What is the graph of $y = -3 \sin(x) + 1$?

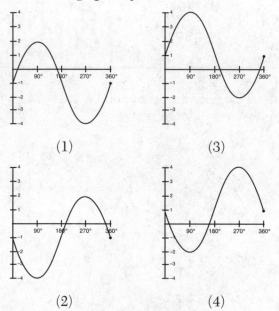

(1)

(3)

(2)

(4)

5. What is the y-coordinate of one of the maximum points of $y = 2 \cos(x) - 1$?

(1) 3   (3) 1

(2) 2   (4) 0

6. What is the period, in degrees, of the graph of $y = 4 \sin\left(\frac{2}{3}x\right) + 1$?

(1) 540°   (3) 270°

(2) 360°   (4) 240°

7. What is the period, in radians, of the graph of
   $y = 4 \cos\left(\frac{\pi}{10}x\right) - 3$?

   (1) 5   (3) 15
   (2) 10   (4) 20

8. Graphically solve the equation $2 \cos(3x) + 3 = 4$ for $0 \leq x < \frac{2\pi}{3}$.

   (1) 0.25 and 1.65   (3) 0.35 and 1.75
   (2) 0.30 and 1.70   (4) 0.40 and 1.80

9. The height above ground of a Ferris wheel car can be modeled
   with the equation $h = -20 \cos\left(\frac{\pi}{15}t\right) + 24$ where h is the height
   in feet and t is the time in seconds.
   (a) How many seconds does it take for the Ferris wheel to
       make a complete revolution?
   (b) What is the maximum height of the Ferris wheel?
   (c) Sketch a graph that shows how the height in the air of a
       Ferris wheel car relates to the amount of time since the
       beginning of the ride?

10. $\cos^2 \theta - \sin^2 \theta$ is equal to which of the following?
    (1) $2 \cos^2 \theta - 1$ (3) -1
    (2) $2 \cos^2 \theta + 1$ (4) 1

Solutions

1. Point B is in quadrant III so the angle is between $180°$ and $270°$. One way to get the solution is to work out

$$180° + \sin^{-1} 0.41 = 204°$$

The correct choice is (**4**).

2. To convert degrees to radians, multiply the number of degrees by $\dfrac{\pi}{180°}$.

$$240° \cdot \frac{\pi}{180°} = \frac{4\pi}{3}$$

The correct choice is (**3**).

3. To convert radians to degrees, multiply the number of radians by $\dfrac{180°}{\pi}$.

$$\frac{\pi}{4} \cdot \frac{180°}{\pi} = 45°$$

The correct choice is (**4**).

4. All the choices have an amplitude of 3. Because of the $+1$, the midline of the solution must be $y = 1$. This eliminates choices (1) and (2). Since the coefficient of the $\sin(x)$ is negative, the sine curve must start by going "down" to the left. This eliminates choice (3).

The correct choice is (**4**).

5. The -1 at the end of the equation means the midline is $y = -1$. The 2 in front of the $\cos(x)$ means the amplitude of the curve is 2 so the maximum points are 2 above the midline or $-1 + 2 = 1$.

The correct choice is (**3**).

6. The formula for period, in degrees, is $\dfrac{360°}{B}$ where B is the coefficient of the x-term. For this example, that is

$$\frac{360°}{\frac{2}{3}} = 360° \cdot \frac{3}{2} = 540°$$

The correct choice is **(1)**.

7. The formula for period, in degrees, is $\dfrac{2\pi}{B}$ where B is the coefficient of the x-term. For this example, that is

$$\frac{2\pi}{\frac{\pi}{10}} = 2\pi \cdot \frac{10}{\pi} = 20$$

The correct choice is **(4)**.

8. Graph $y = 4$ and $y = 2\cos(3x) + 3$ together on the graphing calculator (in radian mode) and find the x-coordinates of the first two intersection points to the right of the y-axis. The intersection points are $(0.35, 4)$ and $(1.75, 4)$.

The correct choice is **(3)**.

9. (a) The length of time that it takes to make one revolution is the period of the curve. This can be calculated with the formula $\dfrac{2\pi}{B}$ where B is the coefficient of t. For this example,

this is $\dfrac{2\pi}{\frac{\pi}{15}} = 2\pi \cdot \dfrac{15}{\pi} = 30$ so the time for one revolution is

30 seconds.

(b) Because of the +24, the midline of the curve is at $y = 24$. The coefficient of the cosine is –20 so the high point is 20 above the midline and $24 + 20 = 44$ feet.

(c) The graph starts at $(0, 44)$ and goes down to $(15, 4)$ and then back up to $(30, 44)$.

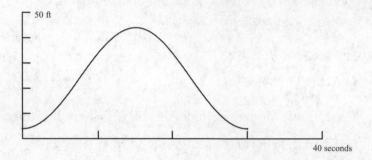

10. One of the most important trig identities is $\sin^2 \theta + \cos^2 \theta = 1$. This can be rearranged as $\sin^2 \theta = 1 - \cos^2 \theta$. Replace the $\sin^2 \theta$ with $1 - \cos^2 \theta$ in the given expression to get

$$\cos^2 \theta - (1 - \cos^2 \theta) = \cos^2 \theta - 1 + \cos^2 \theta = 2\cos^2 \theta - 1$$

The correct choice is **(1)**.

6. SYSTEMS OF EQUATIONS

6.1 TWO EQUATIONS IN TWO UNKNOWNS

A system of two equations in two unknowns is two equations that each have two (or more) variables. A solution to a system of equations is an ordered pair that makes both equations true.

- An example of a system of two equations in two unknowns is

$$x + y = 10$$
$$x - y = 4$$

This system has as its only solution $(7, 3)$ since $7 + 3 = 10$ and $7 - 3 = 4$.

There are two main ways to solve systems of equations with algebra.

The Substitution Method

If one of the variables is already isolated in one of the equations, it is possible to simplify the two equations with two variables into just one equation with one variable.

Example

$$y = 3x - 2$$
$$x + 2y = 24$$

Solution:
Since the y is isolated in the first equation, the y in the second equation can be replaced with $(3x - 2)$.

$$x + 2(3x - 2) = 24$$
$$x + 6x - 4 = 24$$
$$7x - 4 = 24$$
$$\underline{+4 = +4}$$
$$7x = 28$$
$$\frac{7x}{7} = \frac{28}{7}$$
$$x = 4$$

- To get the other number in the ordered pair, substitute the $x = 4$ into the first equation.

$$y = 3 \cdot 4 - 2 = 12 - 2 = 10$$

- The ordered pair that satisfies both equations is (4, 10).

The Elimination Method for Two Equations in Two Unknowns

If the two equations can be combined in a way that *eliminates* one of the variables, the remaining equation can be solved for one part of the ordered pair. Sometimes it requires multiplying both sides of one or both equations by some constant.

Example
Solve the system of equations.

$$2x + 6y = 4$$
$$5x + 3y = 34$$

Solution:
The coefficient of the y in the first equation, 6, is a multiple of the coefficient of the y in the second equation, 3. For the equation that has the smaller coefficient, multiply both sides of the equation by the number that would make that coefficient the opposite of the one from the other equation. For this example, multiply both sides of the second equation by -2.

$$2x + 6y = 4$$
$$-2(5x + 3y) = -2(34)$$

$$
\begin{array}{r}
2x + 6y = 4 \\
+\ -10x - 6y = -68 \\
\hline
-8x = -64
\end{array}
$$

$$\frac{-8x}{-8} = \frac{-64}{-8}$$

$$x = 8$$

$$2 \cdot 8 + 6y = 4$$
$$16 + 6y = 4$$
$$\underline{-16 = -16}$$
$$6y = -12$$
$$\frac{6y}{6} = \frac{-12}{6}$$
$$y = -2$$

The solution is $(8, -2)$.

Example

Solve the system of equations.

$$3x + 4y = 42$$
$$5x - 6y = 32$$

Solution:

For the x-terms, the 5 is not a multiple of 3. For the y-terms, the −6 is not a multiple of the 4. When this happens, both equations must be changed.

You can choose whether you want to change the equation so that either the x-variables will be eliminated or the y-variables will be eliminated.

To eliminate the y-variables, find the least common multiple of the two coefficients. In this case, the least common multiple of 4 and 6 is 12. The goal is to convert one of the y-coefficients into a +12 and the other into a −12. This can be done by multiplying both sides of the first equation by +3 and both sides of the second equation by +2.

$$3(3x + 4y) = 3(42)$$
$$2(5x - 6y) = 2(32)$$

$$9x + 12y = 126$$
$$\underline{+\ 10x - 12y = 64}$$
$$19x = 190$$
$$\frac{19x}{19} = \frac{190}{19}$$
$$x = 10$$

Substitute 10 for x into either of the original equations.

$$3 \cdot 10 + 4y = 42$$
$$30 + 4y = 42$$
$$\underline{-30 = -30}$$
$$4y = 12$$
$$\frac{4y}{4} = \frac{12}{4}$$
$$y = 3$$

The solution is (10, 3).

Infinite Solutions or No Solutions

While doing the systems of equations process, sometimes all the variables get canceled out and you have an equation that has only numbers. If the numbers on either side of the equal sign are not equal, the system has no solution. If the numbers on either side of the equal sign are equal, there are an infinite number of solutions to the system.

Example

How many solutions does this system of equations have?

$$2x + 3y = -10$$
$$-2x - 3y = 12$$

(1) No solutions
(2) One solution
(3) Two solutions
(4) More than two solutions

Solution:

When the two equations are added together it becomes $0 = 2$, which is never true. There are no solutions to the original system of equations. The correct choice is (1).

6.2 THE ELIMINATION METHOD FOR THREE EQUATIONS IN THREE UNKNOWNS

When there are three equations and three variables, the process for solving the system of equations is much longer. It still involves multiplying the equations by constants with the goal of "eliminating" the variables until there is just one variable left.

Example

$$-3x - 7y - 2z = -31$$
$$5x + 4y + 2z = 35$$
$$-x + y - z = -4$$

Solution:

- **Step 1:** Decide which variable to first eliminate.

  The first step in solving a system of three equations with three unknowns is to examine the coefficients of each of the variables to see if the variables for any coefficient are matching numbers or have numbers that are multiples of the other coefficients. Any variable can be eliminated, but it is easier to eliminate a variable when the coefficients contain multiples of other coefficients.

  In this example, the z variables have the coefficients -2, $+2$, and -1. The z will be the variable we try to eliminate.

- **Step 2:** Combine the first and second equation to eliminate a variable.

  Since -2 is the opposite of $+2$, this is like the first example from the previous section. The variable z can be eliminated by adding the two equations. This step may require multiplying both sides of one or both equations by a constant before combining them.

$$
\begin{array}{r}
-3x - 7y - 2z = -31 \\
+ \ 5x + 4y + 2z = 35 \\
\hline
2x - 3y = 4
\end{array}
$$

This equation $2x - 3y = 4$ will be used in step 4.

- **Step 3:** Combine the second and third equations (or the first and third if it is easier) to eliminate a variable.

Since $+2z + -z$ will not eliminate the z, both sides of the second equation must first be multiplied by $+2$ so the z coefficient will become -2, the opposite of the coefficient of the other z, $+2$.

$$5x + 4y + 2z = 35$$
$$2(-x + y - z) = 2(-4)$$

Now the equations can be combined to eliminate the z.

$$\begin{array}{r} 5x + 4y + 2z = 35 \\ + \quad -2x + 2y - 2z = -8 \\ \hline 3x + 6y = 27 \end{array}$$

This equation, together with the one from step 2, is used in step 4.

- **Step 4:** Solve the resulting system of two equations with two unknowns

$$2x - 3y = 4$$
$$3x + 6y = 27$$

Since the $+6$ is a multiple of the -3, multiply both sides of the first equation by $+2$.

$$2(2x - 3y) = 2(4)$$
$$3x + 6y = 27$$

$$\begin{array}{r} 4x - 6y = 8 \\ + \quad 3x + 6y = 27 \\ \hline 7x = 35 \end{array}$$
$$\frac{7x}{7} = \frac{35}{7}$$
$$x = 5$$

Substitute 5 for x into either of the two variable equations.

$$2(5) - 3y = 4$$
$$10 - 3y = 4$$
$$\underline{-10 = -10}$$
$$-3y = -6$$
$$\frac{-3y}{-3} = \frac{-6}{-3}$$
$$y = 2$$

- **Step 5:** Substitute the solution of the two variable system of equations into one of the original equations to find the solution for the third variable.

Any of the original three equations can be used. Since the third equation has the smallest coefficients, it will be simplest.

$$-(5) + 2 - z = -4$$
$$-5 + 2 - z = -4$$
$$-3 - z = -4$$
$$\underline{+3 = +3}$$
$$-z = -1$$
$$\frac{-z}{-1} = \frac{-1}{-1}$$
$$z = 1$$

The solution is the ordered triple (5, 2, 1).

This can be checked by substituting $x = 5$, $y = 2$, and $z = 1$ into each of the three original equations.

$$-3 \cdot 5 - 7 \cdot 2 - 2 \cdot 1 \stackrel{?}{=} -31$$
$$-15 - 14 - 2 \stackrel{?}{=} -31$$
$$-31 \stackrel{\checkmark}{=} -31$$

$$5 \cdot 5 + 4 \cdot 2 + 2 \cdot 1 \stackrel{?}{=} 35$$
$$25 + 8 + 3 \stackrel{?}{=} 35$$
$$35 \stackrel{\checkmark}{=} 35$$

$$-5 + 2 - 1 \stackrel{?}{=} -4$$
$$-4 \stackrel{\checkmark}{=} -4$$

Example

Find the solution to this system of equations

$$-x + 5y + z = -18$$
$$4x - 5y + 3z = 34$$
$$x + 5y - z = -12$$

Solution:

$(4, -3, 1)$

Since the y-variables have coefficients of $+5$, -5, and $+5$, the y should be eliminated. Combining equations 1 and 2 gets $3x + 4z = 16$. Combining equations 2 and 3 gets $5x + 2z = 22$.

Now solve the system

$$3x + 4z = 16$$
$$5x + 2z = 22$$

Multiply both sides of the second equation by -2 so the z coefficient becomes -4.

$$3x + 4z = 16$$
$$\underline{-10x - 4z = -44}$$
$$-7x = -28$$
$$x = 4$$

$$3 \cdot 4 + 4z = 16$$
$$12 + 4z = 16$$
$$\underline{-12 = -12}$$
$$4z = 4$$
$$z = 1$$

$$-4 + 5y + 1 = -18$$
$$-3 + 5y = -18$$
$$\underline{+3 = +3}$$
$$5y = -15$$
$$y = -3$$

The solution is $(4, -3, 1)$.

Practice Exercises

1. Solve this system of equations.
   $y = 2x + 1$
   $3x + 4y = 15$
   (1) $(3, 1)$ (3) $(2, 4)$
   (2) $(4, 2)$ (4) $(1, 3)$

2. Solve this system of equations.
   $y = 3x - 2$
   $7x - 2y = 7$
   (1) $(3, 7)$ (3) $(4, 6)$
   (2) $(7, 3)$ (4) $(6, 4)$

3. Solve this system of equations.
   $3x + 5y = 37$
   $2x - 5y = 8$
   (1) $(9, 2)$ (3) $(8, 3)$
   (2) $(2, 9)$ (4) $(3, 8)$

4. Solve this system of equations.
$$4x + 3y = -11$$
$$-4x + 5y = 35$$
(1) $(5, -3)$ (3) $(-5, 3)$
(2) $(-3, 5)$ (4) $(3, -5)$

5. Solve this system of equations.
$$2x + 3y = 11$$
$$3x - 6y = 6$$
(1) $(4, 1)$ (3) $(2, 3)$
(2) $(1, 4)$ (4) $(3, 2)$

6. Solve this system of equations.
$$5x - 3y = 27$$
$$6x - 7y = 46$$
(1) $(3, -4)$ (3) $(-3, 4)$
(2) $(-4, 3)$ (4) $(4, -3)$

7. How many solutions does this system of equations have?
$$5x + 3y = 10$$
$$10x + 6y = 20$$
(1) No solution   (3) Two solutions
(2) One solution   (4) Infinite solutions

8. If a system of two equations in two unknowns has no solution, what do the graphs of the two lines representing the two equations look like?
(1) They intersect at one point.
(2) They are parallel.
(3) They are the same line.
(4) They intersect at two points.

9. If $x = 2$ and $y = 1$ and $3x + 2y + 4z = 24$, what is the value of z?

10. Solve this system of equations.

$$3x + 2y + z = 15$$
$$2y + z = 6$$
$$z = 4$$

Solutions

1. Since the y is isolated in the first equation, use the substitution method by substituting $2x + 1$ for y in the second equation.

$$3x + 4(2x + 1) = 15$$
$$3x + 8x + 4 = 15$$
$$11x + 4 = 15$$
$$11x = 11$$
$$x = 1$$

To solve for y, replace the x with 1 in the first equation.

$$y = 2(1) + 1 = 2 + 1 = 3$$

The solution is $(1, 3)$.

The correct choice is **(4)**.

2. Since the y is isolated in the first equation, use the substitution method by substituting $3x - 2$ for y in the second equation.

$$6x - 3(3x - 2) = 3$$
$$6x - 9x + 6 = -3$$
$$-3x + 6 = -3$$
$$-3x = -9$$
$$x = 3$$

To solve for y, replace the x with 3 in the first equation.

$$y = 3(3) - 2 = 9 - 2 = 7$$

The solution is $(3, 7)$

The correct choice is **(1)**.

3. Since the coefficient of the y in the first equation is the same number, but opposite sign, as the coefficient of the y in the second equation, the y-terms can be eliminated by adding the two equations.

$$3x + 5y = 37$$
$$\underline{2x - 5y = 8}$$
$$5x = 45$$
$$x = 9$$

Substitute 9 for x into either of the equations and solve for y.

$$3(9) + 5y = 37$$
$$27 + 5y = 37$$
$$5y = 10$$
$$y = 2$$

The solution is (9, 2).

The correct choice is **(1)**.

4. Since the coefficient of the x in the first equation is the same number, but opposite sign, as the coefficient of the x in the second equation, the x-terms can be eliminated by adding the two equations.

$$4x + 3y = -11$$
$$\underline{-4x + 5y = 35}$$
$$8y = 24$$
$$y = 3$$

Substitute 3 for y into either of the equations, and solve for x.

$$4x + 3(3) = -11$$
$$4x + 9 = -11$$
$$4x = -20$$
$$x = -5$$

The solution is (–5, 3).

The correct choice is **(3)**.

5. Since the 6 is a multiple of the 3, multiply both sides of the first equation by 2 so that the coefficient of the modified first equation will be +6. Then the y can be eliminated by adding the modified first equation to the second equation.

$$2(2x + 3y) = 2(11)$$
$$3x - 6y = 6$$

$$4x + 6y = 22$$
$$\underline{3x - 6y = 6}$$
$$7x = 28$$
$$x = 4$$

Substitute 4 for x into either of the original equations to solve for y.

$$2(4) + 3y = 11$$
$$8 + 3y = 11$$
$$3y = 3$$
$$y = 1$$

The solution is (4, 1).

The correct choice is **(1)**.

6. Multiply the top equation by –6 and the bottom by +5 so that the coefficients of the x terms become –30 and +30. Then the modified equations can be added to eliminate the x.

$$-6(5x - 3y) = -6(27)$$
$$5(6x - 7y) = 5(46)$$

$$-30x + 18y = -162$$
$$\underline{30x - 35y = 230}$$
$$-17y = 68$$
$$y = -4$$

Substitute –4 for y into either of the original equations to solve for x.

$$5x - 3(-4) = 27$$
$$5x + 12 = 27$$
$$5x = 15$$
$$x = 3$$

The solution is $(3, -4)$

The correct choice is **(1)**.

7. Multiply the first equation by –2 so that the coefficient on the x becomes –10.

$$-2(5x + 3y) = -2(10)$$
$$10x + 6y = 20$$

$$-10x - 6y = -20$$
$$\underline{10x + 6y = 20}$$
$$0 = 0$$

In trying to just eliminate the x, the y got eliminated, too. Since the equation that remains, $0 = 0$, is true, there are an infinite number of solutions. Had it been something like $0 = 5$, there would be no solutions.

The correct choice is **(4)**.

8. The intersection points of the two lines representing the two equations correspond to the solutions to the system of equations. If there are no solutions, then there are no intersection points. Lines can only have no intersection points if they are parallel.

The correct choice is **(2)**.

9. Substitute 2 for x and 1 for y and solve for z.

$$3(2) + 2(1) + 4z = 24$$
$$6 + 2 + 4z = 24$$
$$8 + 4z = 24$$
$$4z = 16$$
$$z = 4$$

10. Since the third equation says that $z = 4$, substitute 4 for z in the second equation to solve for y.

$$2y + 4 = 6$$
$$2y = 2$$
$$y = 1$$

Substitute 4 for z and 1 for y into the first equation and solve for x.

$$3x + 2(1) + 4 = 15$$
$$3x + 2 + 4 = 15$$
$$3x + 6 = 15$$
$$3x = 9$$
$$x = 3$$

The solution is $(3, 1, 4)$.

7. FUNCTIONS

7.1 COMPOSITE FUNCTIONS

A **function** is a mathematical rule that generally takes a number as an input and then outputs a number. If the function is called f, the notation $f(4) = 11$ means that when the number 4 is put into the function, the number 11 is output from the function.

- Functions are often defined by a formula. For example $f(x) = 2x + 3$ means that when a number is put into the function, a number that is three more than twice that number will come out of the function. So $f(4) = 2 \cdot 4 + 3 = 8 + 3 = 11$.

- Not only can numbers be put into a function, but so can variables, or even other functions. Using the function f defined above, $f(a) = 2a + 3$ and $f(x^2 + 1) = 2(x^2 + 1) + 3$.

- If another function g is defined as $g(x)=3x - 2$, it is possible to create a function $f(g(x))$. By putting the $g(x)$ into the f function, it becomes $f(3x - 2) = 2(3x - 2) + 3$, which can be simplified to $f(3x - 2) = 6x - 4 + 3 = 6x - 1$.

When a function is put into another function, the result is called a **composite function.**

Example 1
If $f(x) = 5x + 2$ and $g(x) = 3x - 1$, what is the value of $f(g(4))$?

Solution:
Since $g(4) = 3 \cdot 4 - 1 = 11, f(g(4))=f(11) = 5 \cdot 11 + 2 = 57.$

Example 2
If $f(x) = 2x + 3$ and $g(x) = 3x - 2$, what is $g(f(x))$ and $f(g(x))$?

Solution:
$g(f(x)) = g(2x + 3) = 3(2x + 3) - 2 = 6x + 9 - 2 = 6x + 7$
$f(g(x)) = f(3x - 2) = 2(3x - 2) + 3 = 6x - 4 + 3 = 6x - 1$

Notice that, in this example, $g(f(x))$ is not equivalent to $f(g(x))$.

7.2 INVERSE FUNCTIONS

If a function has an **inverse function**, the inverse function can take what was output from the original function and turn it back into the original input. An inverse function is a function that "undoes" what some other function has done to a number. The notation for the inverse of a function called f is f^{-1}.

- If the function f is defined by $f(x) = x + 3$, then $f(7) = 7 + 3 = 10$. The inverse of function f is $f^{-1}(x) = x - 3$. When the number 10 is put into the inverse function $f^{-1}(10) = 10 - 3 = 7$, which is what was put into the original function, f, to get the number 10. Whatever the function f does to a number, the function f^{-1} undoes to that number.

Example
If $g(6) = 19$, what is the value of $g^{-1}(19)$?

(1) 6

(2) $\dfrac{1}{6}$

(3) 19

(4) −19

Solution:
Since the function g turned the number 6 into the number 19, then the inverse will turn the 19 back into the 6. With the given information, this is the only value for g^{-1} that can be determined.

The correct choice is (1).

Determining the Inverse Function of a Linear Function

A linear function of the form $f(x) = ax + b$, such as $f(x) = 2x + 3$, has an inverse function that is also a linear function. The process for finding the inverse function is to first rewrite the function, but with the $f(x)$ replaced with an x and the x replaced with an $f^{-1}(x)$.

$$x = 2f^{-1}(x) + 3$$

Then "solve" for $f^{-1}(x)$ by, in this case, subtracting 3 from both sides of the equation and dividing both sides of the equation by 2.

$$x = 2f^{-1}(x) + 3$$
$$\underline{-3 = -3}$$
$$x - 3 = 2f^{-1}(x)$$
$$\frac{x-3}{2} = \frac{2f^{-1}(x)}{2}$$
$$\frac{x-3}{2} = f^{-1}(x)$$

If you pick a number for x like 5, see that $f(5) = 2 \cdot 5 + 3 = 10 + 13$ and that $f^{-1}(13) = \dfrac{13-3}{2} = \dfrac{10}{2} = 5$. So this inverse function undoes what the function did to the number 5.

Graphs of Inverse Functions

If for some function f, $f(3) = 9$, then the point $(3, 9)$ will be on the graph for that function. If $f(3) = 9$, then $f^{-1}(9) = 3$, and the point $(9, 3)$ will be on the graph for the inverse function. In general, if (x, y) is a point on the graph of the original function, then (y, x) will be a graph of the inverse function. In terms of transformations, the point $(9, 3)$ is the reflection over the line $y = x$ of the point $(3, 9)$. When this transformation is done to every point in the graph of the function, the graph of the inverse will be a reflection of the entire graph of the original function over the line $y = x$.

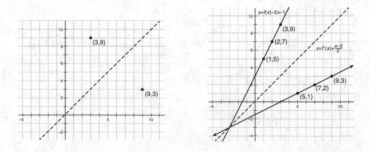

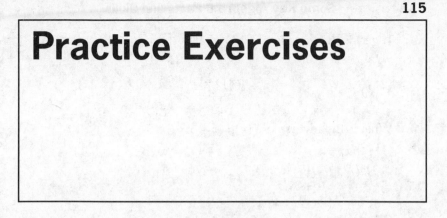

1. If $f(x) = 2x + 3$ and $g(x) = 3x - 4$, what is $f(g(5))$?
   (1) 35   (3) 25
   (2) 30   (4) 20

2. Which of the following could be $f(x)$ and $g(x)$ if
   $f(g(x)) = (3x - 2)^2 + 1$?
   (1) $f(x) = x^2 + 1, g(x) = 3x - 2$
   (2) $f(x) = 3x - 2, g(x) = x^2 + 1$
   (3) $f(x) = x^2 - 1, g(x) = 3x - 2$
   (4) $f(x) = x^2 + 1, g(x) = 3x + 2$

3. If $g(x) = 3x^2 - 2$, what is $g(f(x))$?
   (1) $2[f(x)]^2 + 3$
   (2) $2[f(x)]^2 - 3$
   (3) $3[f(x)]^2 - 2$
   (4) $3[f(x)]^2 + 2$

4. If $g(x)$ is the inverse of $f(x)$, what is the value of $f(g(9))$?

    (1) 3   (3) –9

    (2) $\dfrac{1}{9}$ (4) 9

5. What is the inverse of $f(x) = x - 3$?

    (1) $f^{-1}(x) = x + 3$ (3) $f^{-1}(x) = x - 3$

    (2) $f^{-1}(x) = \dfrac{1}{x+3}$ (4) $f^{-1}(3) = -1$

6. What is the inverse of $f(x) = 2x - 7$?

    (1) $f^{-1}(x) = 2x + 7$ (3) $f^{-1}(x) = \dfrac{x+7}{2}$

    (2) $f^{-1}(x) = \dfrac{1}{2x-7}$ (4) $f^{-1}(x) = \dfrac{x}{2} + 7$

7. If $f(3) = 11$, what must also be true?

    (1) $f^{-1}(11) = 3$ (3) $f^{-1}(3) = \dfrac{1}{11}$

    (2) $f^{-1}(3) = 11$ (4) $f^{-1}(3) = -11$

8. If the graph of $y = f(x)$ is shown below, what is the graph of $y = f^{-1}(x)$?

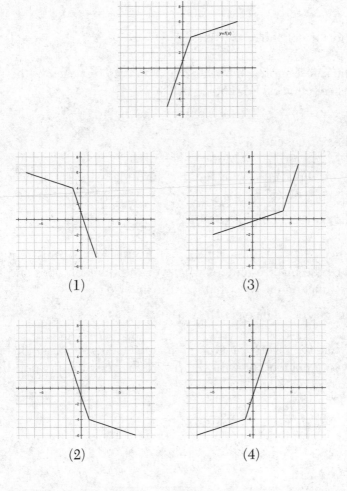

(1)   (3)

(2)   (4)

9. If $f(x) = 5x + 4$, what is $f^{-1}(34)$?

10. If $f(x) = 7x + 3$, what is $f^{-1}(x)$?

Solutions

1. $g(5) = 3(5) - 4 = 15 - 4 = 11$ so

    $$f(g(5)) = f(11) = 2(11) + 3 = 22 + 3 = 25$$

    The correct choice is **(3)**.

2. The safest approach to this question is to test each of the four answer choices. For choice (4),

    $$f(g(x)) = [g(x)]^2 + 1 = (3x - 2)^2 + 1$$

    The correct choice is **(1)**.

3. If you replace the x in the $g(x)$ formula with $f(x)$, it becomes $g(f(x)) = 3[f(x)]^2 - 2$.

    The correct choice is **(3)**.

4. If $g(x) = f^{-1}(x)$, then $f(g(9)) = f(f^{-1}(9))$, which must equal 9 since the inverse undoes whatever the function does and also the function undoes whatever the inverse does.

    The correct choice is **(4)**.

5. To find the inverse of a function, create a new equation like the given equation but with the $f(x)$ replaced with an x and the x replaced with an $f^{-1}(x)$.

    $$x = f^{-1}(x) - 3$$

    Then solve for $f^{-1}(x)$:

    $$\begin{aligned} x &= f^{-1}(x) - 3 \\ +3 &= +3 \\ \hline x + 3 &= f^{-1}(x) \end{aligned}$$

    The correct choice is **(1)**.

6. To find the inverse of a function, create a new equation like the given equation but with the $f(x)$ replaced with an x and the x replaced with an $f^{-1}(x)$.

$$x = 2f^{-1}(x) - 7$$

Then solve for $f^{-1}(x)$:

$$x = 2f^{-1}(x) - 7$$
$$\underline{+7 = +7}$$
$$x + 7 = 2f^{-1}(x)$$
$$\frac{x+7}{2} = \frac{2f^{-1}(x)}{2}$$
$$\frac{x+7}{2} = f^{-1}(x)$$

The correct choice is **(3)**.

7. If $f(3) = 11$, it means that when 3 is put into the function, 11 comes out. So if the inverse function undoes whatever the function does to a number then $f^{-1}(11)$ must be 3.

The correct choice is **(1)**.

8. The graph of the function $f(x)$ has $(-2, -5)$, $(1, 4)$, and $(7, 6)$ as three of the points. The inverse will have three points like these but with the x- and y-coordinates interchanged or $(-5, -2)$, $(4, 1)$ and $(6, 7)$. Of the choices, choice (3) has these three points.

Also the graph of the inverse function is the graph of the original after it has been reflected over the line $y = x$, which is true about choice (3).

The correct choice is **(3)**.

9. The long way is to calculate $f^{-1}(x)$ and then use it to find $f^{-1}(34)$. The short cut for this question is to realize that $f^{-1}(34)$ is the number that would have to go into the original function to make 34 come out or what x value makes $f(x) = 34$. This can be solved with the equation.

$$f(x) = 5x + 4$$
$$34 = 5x + 4$$
$$\underline{-4 = -4}$$
$$30 = 5x$$
$$x = 6$$

10. To find the inverse of a function, create a new equation like the given equation but with the $f(x)$ replaced with an x and the x replaced with an $f^{-1}(x)$.

$$x = 7f^{-1}(x) + 3$$

Then solve for $f^{-1}(x)$:

$$x = 7f^{-1}(x) + 3$$
$$\underline{-3 = -3}$$
$$x - 3 = 7f^{-1}(x)$$
$$\frac{x-3}{7} = \frac{7f^{-1}(x)}{7}$$
$$\frac{x-3}{7} = f^{-1}(x)$$

8. SEQUENCES

8.1 SEQUENCES AND SERIES

A **sequence** is a list of numbers that follow some pattern. An example of a sequence is $a = 1, 4, 9, 16, \ldots$ The individual terms of the sequence are called a_1, a_2, a_3, etc., where the subscript is the term's position on the list.

Explicit Formula for a Sequence

An explicit formula is a type of function into which only positive integers can be input. Sometimes the explicit formula notation resembles function notation like $a(n) = 2n + 3$. An alternative notation is with a subscript, like $a_n = 2n + 3$. By substituting values 1, 2, 3, and 4 for n, the first four terms of the sequence can be determined.

$$a_1 = 2 \cdot 1 + 3 = 5$$
$$a_2 = 2 \cdot 2 + 3 = 7$$
$$a_3 = 2 \cdot 3 + 3 = 9$$
$$a_4 = 2 \cdot 4 + 3 = 11$$

The sequence is $5, 7, 9, 11, \ldots$

Example

What are the first three terms of the sequence defined by equation $a_n = 5 + 3(n - 1)$?

Solution:

Substitute 1, 2, and 3 for n in the formula to get

$$a_1 = 5 + 3(1 - 1) = 5 + 3(0) = 5, \quad a_2 = 5 + 3(2 - 1) = 5 + 3(1) = 8$$

and

$$a_3 = 5 + 3(3 - 1) = 5 + 3(2) = 11$$

Recursive Formula for a Sequence

A recursive formula has two parts. There is the *base case* where the first value (or the first few values) of the sequence are given, and there is the *recursive definition* where a rule for calculating more terms is described in terms of the previous terms.

For example,

$$a_1 = 5$$
$$a_n = 2 + a_{n-1}, \text{ for } n \geq 2$$

To get a_1, there is no calculation since a_1 is given as 5 in the base case part of the definition. For a_2, substitute 2 for n into the recursive part of the definition.

$$a_2 = 2 + a_{2-1} = 2 + a_1 = 2 + 5 = 7$$

Notice that the value of a_1 needed to be known to do this step.
For $n = 3$, $a_3 = 2 + a_{3-1} = 2 + a_2 = 2 + 7 = 9$.
The first three terms are 5, 7, and 9.

Notice that this generates the same sequence as the explicit formula $a_n = 2n + 3$.

Example

What is the value of a_3 for the sequence defined by

$$a_1 = 3$$
$$a_n = n \cdot a_{n-1}, \text{ for } n \geq 2$$

Solution:

$a_1 = 3$ is given.

Substitute 2 for n in the recursive part of the definition to get

$$a_2 = 2 \cdot a_{2-1} = 2 \cdot a_1 = 2 \cdot 3 = 6$$

Substitute 3 for n in the recursive part of the definition to get

$$a_3 = 3 \cdot a_{3-1} = 3 \cdot a_2 = 3 \cdot 6 = 18$$

Arithmetic and Geometric Sequences

A sequence like 5, 7, 9, 11, . . . is known as an *arithmetic sequence* since each term is equal to 2 more than the previous term. The number 2, in this case, is known as the *common difference*, usually denoted by the variable d.

A sequence like 5, 10, 20, 40, . . . is known as a *geometric sequence* since each term is equal to 2 times the previous term. The number 2, in this case, is known as the *common ratio*, usually denoted by the variable r.

The two formulas for the nth term of an arithmetic or geometric sequence are provided on the reference sheet given to you in the Regents booklet.

Arithmetic Sequence	**Geometric Sequence**
$a_n = a_1 + (n-1)d$	$a_n = a_1 r^{n-1}$

Example
What is the 50th term of the sequence 7, 11, 15, 19, . . . ?

Solution:
The explicit formula is $a_n = 7 + 4(n-1)$. Substitute 50 for n to get $a_{50} = 7 + 4(50-1) = 7 + 4(49) = 203$.

Sum of a Finite Geometric Series

A *finite series* is like a finite sequence except all the terms are added together. An example of a finite geometric series is $5 + 10 + 20 + 40 + 80 + 160$.

- The formula for calculating the sum of a finite geometric series is given in the reference sheet in the Regents booklet. The formula is:

$$S_n = \frac{a_1 - a_1 r^n}{1-r}$$

Example

Use the sum of a finite geometric series formula to calculate the sum of $5 + 10 + 20 + 40 + 80 + 160$.

Solution:

To use this formula to calculate $5 + 10 + 20 + 40 + 80 + 160$, substitute 6 for n (since there are 6 terms), 5 for a_1, and 2 for r.

$$S_6 = \frac{5 - 5 \cdot 2^6}{1 - 2} = \frac{5 - 5 \cdot 64}{-1} = \frac{-315}{-1} = 315$$

Practice Exercises

1. What are the first four terms of the sequence defined by $a_1 = 5$, $a_n = 3 + a_{n-1}$ for $n > 1$?
   (1) 5, 8, 11, 14
   (2) 5, 15, 45, 135
   (3) 5, 2, –1, –4
   (4) 5, 3, 1, –1

2. The sequence 3, 12, 48, 192, . . . can be defined by
   (1) $a_1 = 3$, $a_n = 8 + a_{n-1}$ for $n > 1$
   (2) $a_1 = 3$, $a_n = 4a_{n-1}$ for $n > 1$
   (3) $a_1 = 3$, $a_n = 12a_{n-1}$ for $n > 1$
   (4) $a_1 = 3$, $a_n = 12 + a_{n-1}$ for $n > 1$

3. What is the 20th term of the sequence 4, 11, 18, 25, . . . ?
   (1) 130   (3) 144
   (2) 137   (4) 151

4. What position is the number 247 in the sequence
   4, 13, 22, 31, . . . ?
   (1) 26   (3) 28
   (2) 27   (4) 29

5. What is the 10th term of the sequence 4, 20, 100, 500, . . . ?
   (1) 7,812,500   (3) 195,312,500
   (2) 39,062,500   (4) 976,562,500

6. What is the sum of the series

$$2 + 2 \cdot 3 + 2 \cdot 3^2 + 2 \cdot 3^3 + \cdots + 2 \cdot 3^{15}?$$

(1) $\dfrac{2(1-3^{15})}{1-3}$ (3) $\dfrac{2(1+3^{16})}{1+3}$

(2) $\dfrac{2(1+3^{15})}{1+3}$ (4) $\dfrac{2(1-3^{16})}{1-3}$

7. A series is defined as $d_1 = 0$, $d_2 = 1$, $d_n = (n-1)(d_{n-2} + d_{n-1})$ for $n > 2$. What is the value of d_5?
(1) 2   (3) 44
(2) 9   (4) 265

8. What is the sum of the first 20 terms of

$$2 - 6 + 18 - 54 + 162 - 486 + \cdots?$$

(1) –1,743,392,000   (3) –1,743,392,200
(2) –1,743,392,100   (4) –1,743,392,300

9. $\dfrac{1}{2} + \dfrac{1}{4} + \dfrac{1}{8} + \dfrac{1}{16} + \cdots + \dfrac{1}{2^{20}}$ is closest to

(1) 0.99995   (3) 0.9999995
(2) 0.999995   (4) 1

10. For 20 years Journey puts $1,000 into the bank every January 1. The bank pays 3% interest every December 31. If she starts on January 1, 2000, the amount of money she will have on January 2, 2020 is the sequence:

$$1,000 + 1,000(1.05) + 1,000(1.05^2) + \cdots + 1,000(1.05)^{20}$$

How much money is this?

Solutions

1. The first term of the sequence, a_1, is given as 5.
   For $n = 2$: $a_2 + 3 + a_{2-1} = 3 + a_1 = 3 + 5 = 8$
   For $n = 3$: $a_3 + 3 + a_{3-1} = 3 + a_2 = 3 + 8 = 11$
   For $n = 4$: $a_4 + 3 + a_{4-1} = 3 + a_3 = 3 + 11 = 14$
   The first four terms of the sequence are 5, 8, 11, 14.

   The correct choice is (**1**).

2. In this sequence, the first term is 3 and each term is equal to 4 times the previous term. For the base case of the recursive definition, $a_1 = 3$. For the recursive part, the way you define each term as 4 times the previous term is $a_n = 4a_{n-1}$.

   The correct choice is (**2**).

3. This is an arithmetic series with a first term of 4 and a common difference of 7 since each term is 7 more than the previous term. The formula for the nth term of an arithmetic series with a first term of a_1 and a common difference of d is $a_n = a_1 + (n-1)d$. For the 20th term of this sequence, $n = 20$ and $a_{20} = 4 + (20-1)7 = 4 + 19 \cdot 7 = 137$.

   The correct choice is (**2**).

4. This is an arithmetic series with a constant difference of 9 since each term is 9 more than the previous term. To find the position of the number 247 in the sequence, use the formula $a_n = a_1 + (n-1)d$ to solve for n if $a_n = 247$, $a_1 = 4$, and $d = 9$.

$$247 = 4 + (n-1)9$$
$$\underline{-4 = -4}$$
$$243 = (n-1)9$$
$$\frac{243}{9} = \frac{(n-1)9}{9}$$
$$27 = n - 1$$
$$28 = n$$

   The correct choice is (**3**).

5. Since each term is 5 times the previous term, this is a geometric sequence. The nth term of a geometric sequence can be found with the formula $a_n = a_1 r^{n-1}$. If $a_1 = 4$, $r = 5$, and $n = 10$, this becomes $a_{10} = 4 \cdot 5^{10-1} = 4 \cdot 5^9 = 7{,}812{,}500$. Notice that this is not the same as 20^9.

The correct choice is **(1)**.

6. Since each term is 3 times the previous term, this is a geometric series. The formula from the reference sheet given to you on the Regents for the sum of the first n terms of a geometric series is $S_n = \dfrac{a_1 - a_1 r^n}{1 - r}$. For this example, $n = 16$ and not 15 since the first term can be thought of as $2 \cdot 3^0$. Substituting $n = 16$, $a_1 = 2$, and $r = 3$, this becomes

$$S_{16} = \frac{2 - 2 \cdot 3^{16}}{1 - 3} = \frac{2(1 - 3^{16})}{1 - 3}$$

The correct choice is **(4)**.

7. Using the recursive formula for $n = 3$, $n = 4$, and $n = 5$,

$d_3 = (3-1)(d_{3-2} + d_{3-1}) = 2(d_1 + d_2) = 2(0 + 1) = 2$
$d_4 = (4-1)(d_{4-2} + d_{4-1}) = 3(d_2 + d_3) = 3(1 + 2) = 9$
$d_5 = (5-1)(d_{5-2} + d_{5-1}) = 4(d_3 + d_4) = 4(2 + 9) = 44$

The correct choice is **(3)**.

8. Since each term is -3 times the previous term, this is a geometric series with $a_1 = 2$, $r = -3$, and $n = 20$. Using the formula from the reference sheet given to you on the Regents,

$S_n = \dfrac{a_1 - a_1 r^n}{1 - r}$, this becomes

$$S_{20} = \frac{2 - 2 \cdot (-3)^{20}}{1 - (-3)} = -1{,}743{,}392{,}200$$

The correct choice is **(3)**.

9. This is a geometric series with $a_1 = \dfrac{1}{2}$, $r = \dfrac{1}{2}$, and $n = 19$ (and not 20 since the first denominator is 2^1). According to the formula for the sum of the geometric series $S_n = \dfrac{a_1 - a_1 r^n}{1-r}$, for this example $S_{20} = \dfrac{\dfrac{1}{2} - \dfrac{1}{2} \cdot \dfrac{1}{2}^{20}}{1 - \dfrac{1}{2}} = 0.9999990463$. This is a bit closer to 0.9999995 than it is to 1.

The correct choice is **(3)**.

10. This is a geometric series with $a_1 = 1{,}000$, $r = 1.05$, and $n = 21$ (and not 20 since the first term is like $1{,}000 \cdot (1.05)^0$. According to the formula for the sum of a geometric series given in the Regents reference sheet, $S_n = \dfrac{a_1 - a_1 r^n}{1-r}$.

This becomes, for this example,

$$S_{21} = \frac{1{,}000 - 1{,}000 \cdot 1.05^{21}}{1 - 1.05} = 35{,}719.25$$

9. PROBABILITY

9.1 CALCULATING PROBABILITY FROM KNOWN DATA

How likely it is that something happens is measured by a number between 0 and 1. Something certain to happen has a probability of 1, and something that is certain *not* to happen has a probability of 0. Most things have a probability between 0 and 1.

When things are randomly selected from a group, it is often possible to calculate the probability that the selected thing has a certain property.

The formula for calculating probability is

$$P = \frac{\text{number of favorable outcomes}}{\text{number of possible outcomes}}$$

If there is a room with 12 adults and 38 children, for example, and a random person is chosen, the probability that the selected person is a child is $\frac{12}{50} = 0.24$.

When information about a data set is collected on a Venn diagram, it is possible to answer probability questions about that set.

Example

Fifty people are surveyed and asked if they subscribe to Netflix, Amazon Prime, Both, or Neither. The results are collected on the Venn diagram below.

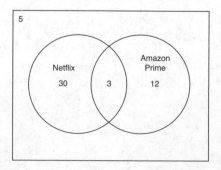

If a person is randomly selected, determine each probability:
1. P(Netflix)
2. P(Amazon Prime)
3. P(Both)
4. P(Neither)
5. P(Netflix or Amazon Prime)
6. P(Netflix if it is known that the person subscribes to Amazon Prime)
7. P(Amazon Prime if it is known that the person subscribes to Netflix)

Solutions:
1. Since $30 + 3 = 33$ people out of 50 subscribe to Netflix, the probability of a randomly chosen person subscribing to Netflix is $\frac{33}{50}$.

2. $12 + 3 = 15$ so $\frac{15}{50}$

3. $\frac{3}{50}$

4. $\frac{5}{50}$

5. This is everyone except the 5 people who subscribe to neither so the answer is $\frac{45}{50}$.

6. There are 15 people who subscribe to Amazon Prime. Of those 15 people, 3 of them subscribe to Netflix (the people who are in the overlap) so the answer is $\frac{3}{15}$.

7. There are 33 people who subscribe to Netflix. Of those 33 people, 3 of them subscribe to Netflix so the answer is $\frac{3}{33}$.

9.2 PROBABILITY OF INDEPENDENT EVENTS

Two events are said to be *independent* if one of the events happening (or not happening) has nothing to do with the other event happening (or not happening).

- An example of two independent events is "It rains in France on December 14, 1987" and "You will ace the Algebra II Regents." One thing has nothing to do with the other. It raining in France on December 14, 1987 does not, in any way, affect your score on the Algebra II Regents.

When events are not independent, they are known as *dependent* events. An example of dependent events is "You study from the Barron's Algebra II review book" and "You will ace the Algebra II Regents." If you study from this Barron's book, you will likely get a higher grade on the Regents than you would if you didn't study from this book.

MATH FACTS

When two events are independent, the probability of both events happening is equal to the product of the probability of each event on its own. In symbols, if A and B are independent events then $P(A \text{ and } B) = P(A) \cdot P(B)$.

For example, if a six-sided die is rolled, the probability of it landing with a 3 facing up is $\frac{1}{6}$ since there is one 3 out of six possible outcomes. If a fair coin is flipped, the probability of it landing with heads facing up is $\frac{1}{2}$ since there is one heads out of two possible outcomes. According to the formula, then, if the die is rolled and the coin is flipped, the probability that the die lands with the 3 facing up *AND* the coin lands with heads facing up is

$$P(3 \text{ and Heads}) = P(3) \cdot P(\text{Heads}) = \frac{1}{6} \cdot \frac{1}{2} = \frac{1}{12}$$

MATH FACTS

For two events, A and B, the probability that A *or* B will happen is the sum of the probabilities of them happening individually *minus* the probability of them both happening.

$$P(A \text{ or } B) = P(A) + P(B) - P(A \text{ and } B)$$

If A and B are independent, this becomes

$$P(A \text{ or } B) = P(A) + P(B) - P(A) \cdot P(B)$$

Example
If the probability that it rains next Tuesday is 0.40 and the probability that the Knicks win the NBA championship is 0.03 and these events are independent, what is the probability that it rains next Tuesday *AND* the Knicks win the NBA championship?

Solution:
Since they are independent events, the probability of both happening is the product of the probabilities of each happening separately. $0.40 \cdot 0.03 = 0.012$.

9.3 PROBABILITY OF DEPENDENT EVENTS

Sometimes it is not obvious whether or not two events are independent or not. There is a way to use math to check. When two events, A and B, are *not* independent, it is no longer true that $P(A \text{ and } B) = P(A) \cdot P(B)$. Instead, it can be quite complicated to determine the probability that both happen. This idea can be used to actually check to see if two events are dependent or independent if we have sufficient information.

MATH FACTS

When checking if two events are independent or not, first multiply $P(A)$ by $P(B)$.
 If $P(A \text{ and } B) = P(A) \cdot P(B)$, A and B are independent events.
 If $P(A \text{ and } B) \neq P(A) \cdot P(B)$, A and B are dependent events.

Example
If $P(A) = 0.30$, $P(B) = 0.40$, and $P(A \text{ and } B) = 0.24$, are events A and B independent or dependent?

Solution:
Since $P(A) \cdot P(B) = 0.30 \cdot 0.40 = 0.12 \neq 0.24$, $P(A) \cdot P(B) \neq P(A \text{ and } B)$ so they are dependent events.

  Using the Netflix vs. Amazon Prime survey data from the diagram below, we can determine if event A = "A person subscribes to Netflix" and B = "A person subscribes to Amazon Prime" are independent.

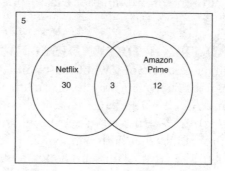

$$P(A) = \frac{33}{50}, P(B) = \frac{15}{50}, \text{ and } P(A \text{ and } B) = \frac{3}{50}$$

Since $P(A) \cdot P(B) = \dfrac{33}{50} \cdot \dfrac{15}{50} = 0.198$ and

$$P(A \text{ and } B) = 0.06, P(A \text{ and } B) \neq P(A) \cdot P(B)$$

so they are not independent events.

Practice Exercises

1. Which two events are independent of one another?
   (1) A = It rains in Nevada on January 1, 1975.
       B = A New York resident goes to the movies on
           May 17, 2016.
   (2) A = It's Valentine's Day.
       B = Flower sales have increased.
   (3) A = A student eats a healthy breakfast.
       B = A student has energy at 10:00 A.M.
   (4) A = A teacher wins the lottery.
       B = The same teacher quits his job.

2. A and B are independent events. If $P(A) = 0.4$ and $P(B) = 0.5$, what is $P(A \text{ and } B)$?
   (1) 0.4
   (2) 0.5
   (3) 0.2
   (4) 0.9

Questions 3, 4, and 5 use the following information.

A fair coin is flipped, and a spinner with the letters A, B, C, D, and E, all equally likely, is spun.

3. What is the probability the coins will land heads and the spinner will land on D?

(1) $\frac{1}{2}$

(3) $\frac{1}{10}$

(2) $\frac{1}{5}$

(4) $\frac{7}{10}$

4. What is the probability that the coin will land on heads or the spinner will land on D?

(1) $\frac{4}{10}$

(3) $\frac{6}{10}$

(2) $\frac{5}{10}$

(4) $\frac{7}{10}$

5. What is the probability that the coin will land on heads given that the spinner landed on D?

(1) $\frac{1}{10}$

(3) $\frac{7}{10}$

(2) $\frac{3}{10}$

(4) $\frac{1}{2}$

6. If the probability that the Rangers win the Stanley Cup is 0.05 and the probability that the Yankees win the World Series is 0.15, what is the probability that both will happen?

(1) 0.20

(3) 0.05

(2) 0.0075

(4) 0.15

7. If $P(A) = 0.4$, $P(B) = 0.6$, and $P(A \text{ and } B) = 0.3$, what conclusion can you make about A and B?
   (1) $P(B|A) = 0.6$
   (2) A and B are independent events.
   (3) A and B are not independent events.
   (4) No conclusion is possible.

8. If $P(A) = 0.7$, $P(B) = 0.6$, and $P(A \text{ and } B) = 0.42$, what conclusion can you make?
   (1) $P(A|B) = 0.6$
   (2) A and B are independent events.
   (3) A and B are not independent events.
   (4) No conclusion can be made.

9. A group of 40 adults are asked if they like classical music, rock music, both, or neither. The results are collected in this Venn diagram.

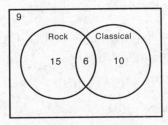

   Based on this data, what conclusion can be made about liking classical music and liking rock music?
   (1) $P(B) \neq P(B|A)$ (A = rock and B = classical)
   (2) They are not independent of each other.
   (3) They are independent of each other.
   (4) No conclusion can be made.

10. 104 students were polled about whether they like soccer, basketball, both, or neither. Based on this data determine if liking basketball and liking soccer are independent events?

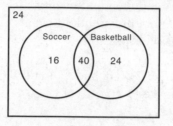

Solutions

1. In choices (2), (3), and (4) there is a cause-and-effect relationship. If the event A happens for those choices, then event B is more likely to happen than it would be if event A did not happen.

The correct choice is **(1)**.

2. When events are independent, $P(A \text{ and } B) = P(A) \cdot P(B)$. Since these events are independent,

$$P(A \text{ and } B) = P(A) \cdot P(B) = 0.4 \cdot 0.5 = 0.2$$

The correct choice is **(3)**.

3. Since there are two sides to the coin and only one of them is heads, $P(\text{Heads}) = \dfrac{1}{2}$. Since there are five different letters on the spinner and only one of them is D, $P(D) = \dfrac{1}{5}$.

The coin and spinner are independent of one another so the formula $P(A \text{ and } B) = P(A) \cdot P(B)$ can be used. For this example,

$$P(\text{Heads and D}) = P(\text{Heads}) \cdot P(D) = \frac{1}{2} \cdot \frac{1}{5} = \frac{1}{10}$$

The correct choice is **(3)**.

4. Since there are two sides to the coin and only one of them is heads, $P(\text{Heads}) = \frac{1}{2}$. Since there are five different letters on the spinner and only one of them is D, $P(D) = \frac{1}{5}$.

The coin and spinner are independent of one another so the formula $P(A \text{ and } B) = P(A) + P(B) - P(A) \cdot P(B)$ can be used. For this example

$$P(\text{Heads or D}) = P(\text{Heads}) + P(D) - P(\text{Heads}) \cdot P(D)$$

$$= \frac{1}{2} + \frac{1}{5} - \frac{1}{2} \cdot \frac{1}{5} = \frac{5}{10} + \frac{2}{10} - \frac{1}{10} = \frac{6}{10}$$

The correct choice is **(3)**.

5. Since the events are independent, the probability of the coin landing on heads is not affected by whether or not the spinner lands on D. If it is given that the spinner landed on D, the probability of the coin landing on heads is still $\frac{1}{2}$.

The correct choice is **(4)**.

6. These are independent events since one happening has no affect on the other happening. For independent events, the formula $P(A \text{ and } B) = P(A) \cdot P(B)$ can be used.

For this example,

$$P(\text{Stanley Cup and World Series})$$
$$= P(\text{Stanley Cup}) \cdot P(\text{World Series})$$
$$= 0.05 \cdot 0.15 = 0.0075$$

The correct choice is **(2)**.

7. When events are independent, $P(A \text{ and } B) = P(A) \cdot P(B)$. In this example, $P(A) \cdot P(B) = 0.4 \cdot 0.6 = 0.24$, but it is given that $P(A \text{ and } B) = 0.3 \neq 0.24$. So these are not independent events.

   The correct choice is **(3)**.

8. When events are independent, $P(A \text{ and } B) = P(A) \cdot P(B)$. In this example, $P(A) \cdot P(B) = 0.7 \cdot 0.6 = 0.42$, and it is given that $P(A \text{ and } B) = 0.42$. So these are independent events.

   The correct choice is **(2)**.

9. The events are independent if

   $$P(\text{rock and classical}) = P(\text{rock}) \cdot P(\text{classical})$$

   Since there are 40 people and 6 of them like both,

   $$P(\text{rock and classical}) = \frac{6}{40} = 0.15$$

   From the Venn diagram, there are $15 + 6 = 21$ people who like rock and $10 + 6 = 16$ people who like classical so

   $$P(\text{rock}) = \frac{21}{40} \text{ and } P(\text{classical}) = \frac{16}{40}$$

   This means that $P(\text{rock}) \cdot P(\text{classical}) = \frac{21}{40} \cdot \frac{16}{40} = 0.21 \neq 0.15$

   so these are not independent events.

   The correct choice is **(2)**.

10.

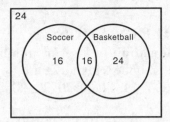

To see if liking soccer and liking basketball are independent, check to see if

$P(\text{soccer}) \cdot P(\text{basketball}) = P(\text{soccer and basketball})$

Because 16 + 16 = 32, 32 out of 40 people like soccer and $P(\text{soccer}) = \dfrac{32}{80}$.

Because 24 + 16 = 40, 40 out of 80 people like basketball and $P(\text{basketball}) = \dfrac{40}{80}$.

There are 16 out of 80 people who like both sports so

$$P(\text{soccer and basketball}) = \frac{16}{80} = 0.2$$

Multiplying $P(\text{soccer}) \cdot P(\text{basketball}) = \dfrac{32}{80} \cdot \dfrac{40}{80} = 0.2$.

Because $P(\text{soccer}) \cdot P(\text{basketball}) = P(\text{soccer and basketball})$, these are all independent events.

10. NORMAL DISTRIBUTION

10.1 STANDARD DEVIATION

The **standard deviation** of a set of numbers is a measure of how "spread out" the numbers in the set are.

The 14 numbers 1, 7, 14, 14, 20, 20, 20, 20, 20, 20, 26, 26, 33, 39 have a mean of 20.

The 14 numbers 18, 19, 19, 19, 20, 20, 20, 20, 20, 20, 21, 21, 21, 22 also have a mean of 20.

Though these two data sets have the same mean, the numbers in the second data set are less spread out. In statistics, we say that the second set of numbers has a lower standard deviation. The standard deviation for the second set is 1, whereas the standard deviation for the first set is 9.3. To have a standard deviation of zero, all the numbers in the data set would have to be the same number.

Calculating Standard Deviation with a Calculator

The standard deviation of a set of numbers can be quickly calculated with a graphing calculator.

* Here is how to calculate the standard deviation for the 14 numbers in the first data set 1, 7, 14, 14, 20, 20, 20, 20, 20, 20, 26, 26, 33, 39.

For the TI-84:

Press [STAT] [1] to get to the list editor. Go to the L1 cell and press [CLEAR] [ENTER]. Enter the 14 scores into L1. Press [STAT], move to CALC, press [1]. Make List: L1 and leave FreqList: blank. Go to Calculate, and press [ENTER]. The first value, 20, is the mean. The fifth value, next to the σx= is the standard deviation. The standard deviation for this data set is 9.3.

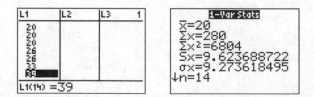

For the TI-Nspire:

From the home screen, select the Add Lists & Spreadsheet icon. Name column A x and fill in cells A1 through A14 with the 14 scores. Press [menu] [4] [1] [1] for One-Variable Statistics. Set the Number of Lists to 1. Select the [OK] button. Set the X1 List to x. In cell C2 will be the mean, 20. In cell C6 will be the standard deviation next to the σx=. The standard deviation for this data set is 9.3.

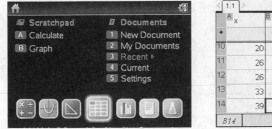

Standard Deviation for Data Sets with Repeated Values

Since there are repeated values in the data set, the values can be listed in a two-column chart. The first column is for the values, and the second column is for how many of that value is in the data set. For this set, it would look like this.

x_i	f_i
1	1
7	1
14	2
20	6
26	2
33	1
39	1

For the TI-84:

Press [STAT] [1] and enter the numbers 1, 7, 14, 20, 26, 33, and 39 into L1 and enter the numbers 1, 1, 2, 6, 2, 1, and 1 into L2. Press [STAT], move to CALC, and press [1]. In List: put "L1" by pressing [2ND] [1]. In FreqList: put "L2" by pressing [2ND] [2]. Move to Calculate and press [ENTER].

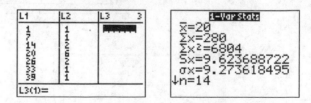

For the TI-Nspire:

From the home screen, select the Add Lists & Spreadsheet icon. Name column A x and fill in cells A1 through A7 with the 7 scores. Name column B y and fill in cells B1 through B7 with the 7 frequencies. Press [menu] [4] [1] [1] for One-Variable Statistics. Set the number of lists to 1. Select the [OK] button. Set the X1 List to x and the Frequency List to y. In cell C2 will be the mean, 20. In cell C6 will be the standard deviation next to the σx=. The standard deviation for this data set is 9.3.

- **z-scores**

In the data set from before with a mean of 60 and a standard deviation of 4, the number 64 was one standard deviation above the mean since it is equal to the mean plus the standard deviation, $60 + 4 = 64$. The number 68 was two standard deviations above the mean since it is equal to the mean plus two times the standard deviation $60 + 2 \cdot 4 = 68$.

In statistics, we say that a number that is one standard deviation above the mean has a *z-score* of 1, whereas a number that is two standard deviations above the mean has a *z-score* of 2.

MATH FACTS

The *z-score* of a number is how many standard deviations above the mean that number is. If the number is greater than the mean, the *z-score* will be positive. If it is less than the mean, the *z-score* will be negative. A formula for calculating *z-score* is

$$z = \frac{\text{number} - \text{mean}}{\text{standard deviation}}$$

Example
If the mean is 60 and the standard deviation is 4, what is the *z-score* of the number 52?

Solution:
Since 52 is less than 60, the *z-score* will be negative. For this example, it is possible to subtract the standard deviation, 4, from 60 to get to 56, which would have a *z-score* of −1. Then subtract it again to get to 52, which has a *z-score* of −2.

This can also be calculated with the formula

$$z = \frac{\text{number} - \text{mean}}{\text{standard deviation}} = \frac{52 - 60}{4} = \frac{-8}{4} = -2$$

10.2 NORMAL DISTRIBUTION

> ### KEY IDEAS
>
> The dot plot of real-world data often has the shape of a *bell curve* also known as a *normal distribution curve*. When this happens, it is possible to answer certain questions about the data that you would not be able to answer had the shape of the dot plot not been a bell curve.

The Normal Curve

Below is a dot plot with 100 data points representing the weights of 100 children. The mean value is approximately 60 and the standard deviation is approximately 4.

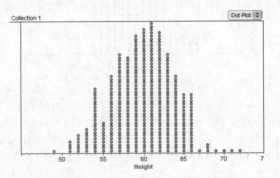

Since the mean is 60 and the standard deviation is 4, a score of 64 is said to be *one standard deviation above the mean*. A score of 68 is said to be *two standard deviations above the mean*. A score of 56 is said to be *one standard deviation below the mean*. A score of 52 is said to be *two standard deviations below the mean*.

Using the Calculator to Find the Percentage of Numbers Between Two Numbers If the Mean and Standard Deviation Are Known

- If the mean and the standard deviation of a data set are known, the graphing calculator can quickly determine what percent of the numbers in the data set will be between any two values.

The heights of 400 people are measured; the mean is approximately 60, and the standard deviation is approximately 4.

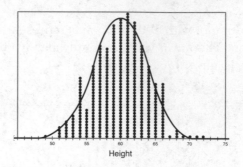

When the data has a normal distribution, the calculator can be used to determine the percent of numbers between two values. To find what percent of people in the above example are between 57.5 inches and 62 inches, do the following:

For the TI-84:

Press [2ND] [VARS] [2] for normalcdf. Fill in the fields with 57.5 for lower, 62 for upper, 60 for μ. Then select Paste.

```
DISTR DRAW          normalcdf          normalcdf(57.5,▶
1:normalpdf(        lower:57.5              .425477
2:normalcdf(        upper:62
3:invNorm(          μ:60
4:invT(             σ:4
5:tpdf(             Paste
6:tcdf(
7↓χ²pdf(
```

For the TI-Nspire:

From the home screen, press [A]. Press [menu] [6] [5] [2] for Normal Cdf. Set the Lower Bound to 57.5, the Upper Bound to 62, the mean (μ) to 60, and the standard deviation (σ) to 4.

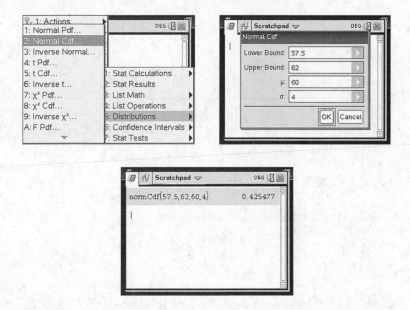

The percent is approximately 43%, which is approximately $0.43 \cdot 400 = 172$ people between 57.5 inches and 62 inches tall.

For examples when there is no lower bound, make the lower bound a large negative number like −9999. If there is no upper bound, make the upper bound a large positive number like 9999.

Practice Exercises

1. Each of these data sets has a mean of 5. Which has the smallest standard deviation?
   (1) 1, 2, 5, 7, 10   (3) 4, 5, 5, 5, 6
   (2) 2, 3, 5, 5, 10   (4) 3, 4, 5, 5, 8

2. What is the standard deviation of the data set 16, 19, 14, 12, 10, 15, 15, 16?
   (1) 2.5   (3) 2.7
   (2) 2.6   (4) 2.8

3. What is the standard deviation for this data set?

x_i	f_i
18	4
19	7
20	2
21	3
22	6

   (1) 1.41   (3) 1.54
   (2) 1.51   (4) 1.61

4. If the mean of a data set is 60 and the standard deviation is 8, what is the *z-score* for the number 72?
   (1) 0.67   (3) 1.25
   (2) 1.00   (4) 1.50

5. How many of these 8 numbers are more than two standard deviations from the mean?

   18, 40, 44, 45, 48, 50, 50, 68

   (1) 1   (3) 3
   (2) 2   (4) 0

6. If the mean of a set of normally distributed 200 numbers is 70 and the standard deviation is 6, approximately what percent of the numbers will be between 64 and 76?
   (1) 68%   (3) 76%
   (2) 70%   (4) 80%

7. The weights of 400 children are collected, and the data are normally distributed. What percent of the children were between 46 and 52 pounds if the mean was 50 pounds and the standard deviation was 3?
   (1) 57%   (3) 63%
   (2) 60%   (4) 66%

8. On the Algebra II Regents, the mean score was 75 and the standard deviation was 9. If the data are normally distributed, what percent of the scores were over 90?
   (1) 20%   (3) 10%
   (2) 15%   (4) 5%

9. The amount of time a group of 300 adults exercised per week was collected. If the data were normally distributed, what percent of the people exercised less than 20 minutes a week if the mean was 40 minutes and the standard deviation was 15?

(1) 9%   (3) 13%
(2) 11%   (4) 15%

10. In a set of 500 samples, the mean is 90 and the standard deviation is 17. If the data are normally distributed, how may of the 500 are expected to have a value between 93 and 101?

(1) 82   (3) 86
(2) 84   (4) 88

Solutions

1. When the numbers are all very close together, the standard deviation is close to 0. The smallest possible standard deviation is 0 when all the numbers are equal to the mean, as for the data set 5, 5, 5, 5, 5. Of the four choices, choice (3) has three of the numbers equal to the mean, and the other two are just one away from the mean so it has the smallest standard deviation.

The correct choice is **(3)**.

2. Enter the eight numbers into the graphing calculator and do the 1-Variable Stats. The calculator will output a bunch of values. Find the number next to σx=, which, in this case, is 2.54666, which is closest to 2.5.

The correct choice is **(1)**.

3. In the graphing calculator, enter the values 18, 19, 20, 21, 22 into the first column and the numbers 4, 7, 2, 3, 6 into the second column. Follow the instructions for lists with frequencies given in this section. Find the number next to σx=, which, in this case, is 1.507556, which is closest to 1.51.

The correct choice is **(2)**.

4. The *z-score* can be calculated with the formula

$$z = \frac{\text{number} - \text{mean}}{\text{standard deviation}}$$

For this example, $z = \frac{72-60}{8} = \frac{12}{8} = 1.5$. The number 72 has a *z-score* of 1.5 so it is 1.5 standard deviations above the mean.

The correct choice is **(4)**.

5. Using the graphing calculator, the mean for these 8 numbers is 45.375, and the standard deviation is 12.95. Numbers more than two standard deviations above the mean are over $45.375 + 2 \cdot 12.95 = 71.275$ or under $45.375 - 2 \cdot 12.95 = 19.475$. In the list of 8 numbers, there are no numbers above 71.275 and one number, 18, below 19.475. Only one number, 18, is more than two standard deviations from the mean.

The correct choice is **(1)**.

6. Using the normal cdf function on the calculator, enter a mean of 70, a standard deviation of 6, a lower bound of 64, and an upper bound of 76. The output is 0.68, which means that in a normal distribution with a mean of 70 and a standard deviation of 6, about 68% of the values are expected to be between 64 and 76.

The correct choice is **(1)**.

7. Using the normal cdf function on the calculator, enter a mean of 50, a standard deviation of 3, a lower bound of 46, and an upper bound of 52. The output is 0.6563, which means that in a normal distribution with a mean of 50 and a standard deviation of 3, about 66% of the values are expected to be between 46 and 52.

The correct choice is **(4)**.

8. Using the normal cdf function on the calculator, enter a mean of 75, a standard deviation of 9, a lower bound of 90, and, since there is no upper bound, an upper bound of 999. The output is 0.04779, which means that in a normal distribution with a mean of 75 and a standard deviation of 9, about 5% of the values are expected to be greater than 90.

    The correct choice is **(4)**.

9. Using the normal cdf function on the calculator, enter a mean of 40, a standard deviation of 15, an upper bound of 20, and, since there is no lower bound, a lower bound of –999. The output is 0.09121, which means that in a normal distribution with a mean of 40 and a standard deviation of 15, about 9% of the values are expected to be less than 20.

    The correct choice is **(1)**.

10. Using the normal cdf function on the calculator, enter a mean of 90, a standard deviation of 17, a lower bound of 93, and an upper bound of 101. The output is 0.1711, which means that in a normal distribution with a mean of 90 and a standard deviation of 17, about 17.12% of the values are expected to be between 93 and 101. Since there are 500 samples, find 17.12% of 500, which is $0.1712 \cdot 500 = 85.6 \approx 86$.

    The correct choice is **(3)**.

11. INFERENTIAL STATISTICS

11.1 DIFFERENT TYPES OF STATISTICAL STUDIES

The three types of statistical studies you should be familiar with are surveys, observational studies, and experimental studies.

- A **survey** is a bunch of questions that are asked about the subjects you are studying.
- In an **observational study**, the subjects in the study get observed and data about them is recorded, but the observer does not try to intervene and control anything about the subjects.
- In an **experimental study**, the observer separates the subjects into groups and requires the different groups to do different things.

If Barron's wanted to know how well the Algebra II red review book is working, they could use any of these three methods.

If they wanted to use a survey they could send out letters to a random list of students and ask two questions: (1) Did you study with the Algebra II red review book? (2) What did you get on the Algebra II Regents? They would want to be careful to try to choose a random group of students and not introduce bias by sending to people more or less likely to have used the book.

If they wanted to do an observational study, they could monitor a group of people and observe whether or not they studied with the Algebra II red review book and then observe them taking the Regents and reacting to their eventual score.

If they wanted to do an experimental study, they could pick 100 random students and give 50 of them the Algebra II red review book and tell the other students not to use the Algebra II red review book. Data would then be collected to see how the students in the different groups performed. It is important that the groups are randomly selected so that there is no bias to distort the results.

11.2 ANALYZING COMPUTER GENERATED HISTOGRAMS

There are three situations where you might see a computer-generated histogram that looks something like an approximate bell curve.

When asked to analyze this, the general guideline is that if the thing you are being asked about is either not represented on the histogram at all or represented by an extremely small bar, then that event is not very likely.

On the other hand, if the thing you are being asked about is represented by a bar that is not small, then it is at least somewhat likely that the thing will happen.

Sometimes in addition to the picture of the histogram, additional statistics are given like the mean and the standard deviation. With these numbers, it is possible to tell if the thing you are asked about is likely or unlikely, even without looking at the histogram. If the thing you are asked about is more than two standard deviations from the mean, it is very unlikely. If it is less than two standard deviations from the mean, then it is at least somewhat likely that the thing will happen.

Example
Thirty students in a focus group are asked to give their opinion on a new style of backpack. Only 9 of the 30 students polled, or 30%, liked the new backpack. The manufacturer wants to know if is likely that out of the entire population, 40% would like the backpack.

A computer is used to generate 100 random sets of 30 students from an imaginary group of students from which exactly 40% of them do like the backpack and the percent who like the backpack for each sample set is graphed on a dot plot or histogram and the mean and the standard deviation of those numbers is calculated.

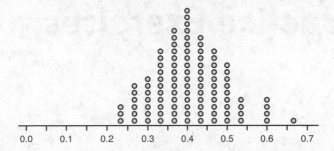

The mean value was 0.404 and the standard deviation was 0.089.

Is it likely that 40% of the entire population actually would like the backpack even though only 30% of the 30-person sample liked it?

Solution:
On the histogram it can be seen that the bars representing 30% or less were not particularly small. This means that even if the percent of the full sample who liked the backpack was 40%, it would be somewhat likely for samples of 30 to have 30% or less liking them.

This could also be done without looking at the histogram, but instead seeing if 30% was within two standard deviations of the mean.

For this example, $0.404 - 2 \cdot 0.089 = 0.226$ and $0.404 + 2 \cdot 0.089 = 0.582$. Since 30% is between 22.6% and 58.2%, it is likely for 30% of a random sample of 30 people to like the backpack even if really 40% of the entire population liked it.

Practice Exercises

1. You want to find out what type of movie is the favorite among the residents of New York City. If you want to do this by conducting a survey, what are some ways that you can reduce bias?

2. You want to study how the amount of sleep a student gets the night before the Regents relates to their Regents score. How can this be studied as an observational study? How can this be studied as an experiment?

3. There are 500 10th graders at Regentsville High School. You want to do an experiment where 20 students study for the Regents with just the Red Barron's book and 20 other students study for the Regents with the Blue and the Red Barron's books. How can the groups be chosen to reduce the chance of skewed data through bias?

4. A survey is conducted to learn what subjects students enjoy most at school. One question reads: "What is your most boring subject?" Could the wording of this question skew the results of the survey? Explain.

5. Using a population as of 10,000 fish in a lake, an ecologist takes 500 samples of size 50. The average for all these samplings is 0.24 with a standard deviation of 0.06.

This is a histogram of the sampling distribution of the sample proportion.

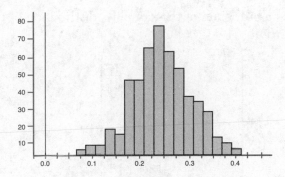

Using this data, with a 95% confidence interval, we can determine the percent of fish in the lake that are cod is which of the following?

(1) Between 0.12 and 0.36.
(2) Between 0.14 and 0.34.
(3) Between 0.20 and 0.28.
(4) Exactly 0.24.

6. A town has 300,000 households and a statistician at a phone company wants to know what percent of them have a landline. She takes 300 samplings, each with size 100 and calculates the percent of households with a landline for each of the samplings. The average of all the means was 0.73 and the standard deviation was 0.05.

This is a histogram of the sampling distribution of the sample proportion.

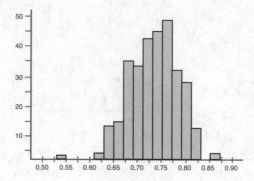

Using this data, with a 95% confidence interval, we can determine that the percent of homes in the town that have a land line is

(1) between 0.63 and 0.83.
(2) between 0.68 and 0.78.
(3) between 0.70 and 0.76.
(4) exactly 0.73.

7. A sports doctor conducts an observational study to learn the average amount of time that 3,000 swimmers in the town can hold their breath underwater. He creates 150 random samplings of 60 people per sampling where for each sampling the average amount of time that the swimmers can hold their breath underwater was calculated. The average of all the means of all the samplings was 72.7, and the standard deviation was 0.92.

This is a histogram of the sampling distribution of the sample mean.

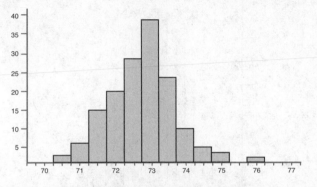

Based on this data, with a 95% confidence interval, the researchers can determine that the actual average amount of time the entire population can hold its breath under water is

(1) exactly 72.7.
(2) between 72 and 73.4.
(3) between 71.28 and 73.12.
(4) between 70.86 and 74.54.

8. A data specialist at a social media website wants to estimate the average age of the 20,000 people attending a Taylor Swift concert. For 150 random samplings of 80 people per sample, the average age of the 80 person sample was calculated. The average of all the means of all the samplings was 25.5, and the standard deviation was 1.5.

This is a histogram of the sampling distribution of the sample mean.

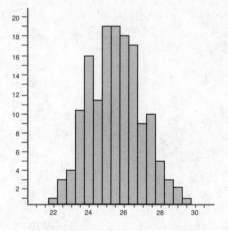

Based on this data, with a 95% confidence interval, the researchers can determine that the average age of the entire 20,000 person population is

(1) exactly 25.5.
(2) between 22.5 and 28.5.
(3) between 23.5 and 27.5.
(4) between 24.5 and 26.5.

9. Amy Hogan rolls a six-sided die 12 times and gets a 6 on four of the rolls. She wants to examine if this might suggest that the die is defective since, on average, the number of expected sixes in 12 rolls is two. She uses a computer to simulate 500 times of rolling a fair die 12 times. A histogram of the results is shown below.

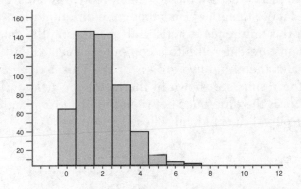

Based on this histogram, Amy concludes that,

(1) yes, the die is defective because two 6s is most common.

(2) yes, the die is defective because four 6s is not common.

(3) no, the die is not necessarily defective because two 6s is not very common.

(4) no, the die is not necessarily defective because four 6s is fairly common.

10. A new plant food is developed that the inventors claim helps plants grow faster. They cite an experiment they did where out of 40 randomly chosen plants, 20 of them were randomly chosen to get the new plant food while the other 20 plants got a different kind of plant food.

The 20 plants that got the new plant food grew an average of 2 inches more than the plants that got a different plant food. To test if this difference is statistically significant, the data for the 40 plants are entered into a computer which, then randomly separates them into two groups of 20. This is done 50 times, and the results are shown in the histogram below where the number is the difference between the first group and the second group of the randomly chosen groupings.

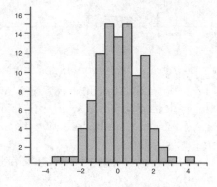

Based on this data, can you conclude that the new plant food causes plants to grow faster?

(1) No, the difference is not statistically significant. +2 is uncommon.

(2) No, the difference is not statistically significant. +2 is fairly common.

(3) Yes, the difference is statistically significant. +2 is uncommon.

(4) Yes, the difference is statistically significant. +2 is fairly common.

Solutions

1. Bias in a survey happens when people more likely to answer one way are more frequently surveyed than people more likely to answer the other way. To make sure this doesn't happen in the movie survey in New York City, be sure to not find people for the survey who live in the same neighborhood or are about the same age.

   The best way to reduce bias is to take a list of all the possible subjects for the survey and have a computer randomly select people for the survey from the list.

2. For an observational survey, you would observe who gets what amount of sleep and what their Regents score is. You would not, in any way, control who gets what amount of sleep.

   For an experiment, you would get a random group of people and then randomly separate those people into groups. For one group, you would tell them to get more than 10 hours of sleep. For another group, you would tell them to get between 7 and 10 hours of sleep, and for a third group you would tell them to get less than 7 hours of sleep. Later, you would record what scores the people from the different groups got on the test and look at that data for patterns or trends.

3. The best way to reduce bias is to use a computer with a random number generator to pick who uses which book. You do not want to let the participants select which group they are in since that could distort the results.

4. The wording of a question could introduce bias. The word "boring" has connotations that could influence the person taking the survey to answer differently than they might with a more neutrally worded question.

5. According to the histogram, the percent of cod in the lake is somewhere between approximately 10% and 40% since anything smaller than 10% or bigger than 40% does not have a bar on it. Since they give the mean and the standard deviation, you can approximate the true percent by calculating mean $- 2 \cdot$ standard deviation and mean $+ 2 \cdot$ standard deviation. For this example, that is

$$0.24 - 2 \cdot 0.06 = 0.12 \text{ and } 0.24 + 2 \cdot 0.06 = 0.36$$

The correct choice is **(1)**.

6. According to the histogram, the percent of people having a land line is somewhere between approximately 55% and 85% since anything smaller than 55% or bigger than 85% does not have a bar on it. Since they give the mean and the standard deviation, you can approximate the true percent by calculating mean $- 2 \cdot$ standard deviation and mean $+ 2 \cdot$ standard deviation. For this example, that is

$$0.73 - 2 \cdot 0.05 = 0.63 \text{ and } 0.73 + 2 \cdot 0.05 = 0.83$$

The correct choice is **(1)**.

7. According to the histogram, the amount of time people can hold their breath is somewhere between 70 and 76 seconds since anything less than 70 or greater than 76 does not have a bar on it. Since they give the mean and the standard deviation, you can approximate the true percent by calculating mean $- 2 \cdot$ standard deviation and mean $+ 2 \cdot$ standard deviation. For this example, that is

$$72.7 - 2 \cdot 0.92 = 70.86 \text{ and } 72.7 + 2 \cdot 0.92 = 74.54$$

The correct choice is **(4)**.

8. According to the histogram, the average age of the people at the concert is somewhere between 22 and 30 since anything less than 22 or greater than 30 does not have a bar on it. Since they give the mean and the standard deviation, you can approximate the true percent by calculating mean − 2 · standard deviation and mean + 2 · standard deviation. For this example, that is

$$25.5 - 2 \cdot 1.5 = 22.5 \text{ and } 22.5 + 2 \cdot 1.5 = 28.5$$

   The correct choice is (**2**).

9. Based on the histogram, it is most likely to get one or two 6s when a fair die is rolled 12 times. However, notice that there is a bar for four 6s and that that bar is more on the medium to small side, but would not be considered "tiny." Because of this, it can't be concluded that the die is defective.

   The correct choice is (**4**).

10. According to the histogram, randomly separating all the plants into two groups usually makes it so there is 0 difference between the two groups. However, notice that there is a visible bar over the 2 that is not relatively "tiny" compared to the others. This means that a difference of 2 is fairly common so we would say that the difference is not statistically significant.

    The correct choice is (**2**).

Glossary of Terms

Algebraic identity Two expressions usually involving polynomials that can be shown to be equal to each another.

Amplitude The distance between the middle of sine or cosine curve to the top of that sine or cosine curve. When the equation is $y = a \sin(bx) + d$ or $y = a \cos(bx) + d$, the amplitude is a.

Arithmetic sequence A sequence of numbers like 3, 7, 11, 15, . . . , where each number is the same amount more (or less) than the previous number. That difference is called d. In this case, $d = 4$.

Arithmetic series The sum of the terms in an arithmetic sequence. $3 + 7 + 11 + 15$ is an arithmetic series.

Asymptote A line that a curve gets closer and closer to without ever reaching.

Base In an exponential expression, the base is the number being raised to a power. In the expression $5 \cdot 2^x$, 2 is the base. In a logarithmic expression, the base is the small subscript number after the word "log." In the expression $\log_3 81 = 4$, the 3 is the base.

Binomial A polynomial with two terms. The expression $2x + 5$ is a binomial.

Coefficient The number that gets multiplied by a variable in a mathematical expresssion. In the expression $5x^2$, the coefficient of the x^2 term is 5.

Complex number A number of the form $a + bi$ where a and b are real numbers. $5 + 3i$ is a complex number.

Composite function When the expression that defines one function is put into another function. If $f(x) = 3x$ and $g(x) = x^2$, then $g(f(x)) = (3x)^2$ is a composite function.

Compound interest When money in a bank gets interest on an initial deposit and then later gets more interest based on the original deposit and on the interest already earned.

Constant A number; the term in a polynomial that has no variable part. In the polynomial $2x + 3$, the constant is 3.

Cosine The ratio between the adjacent side and hypotenuse of a right triangle. Also the x-coordinate of a point on the unit circle.

Cubic equation An equation of the form $ax^3 + bx^2 + cx + d = 0$.

Cubic function A function of the form $f(x) = ax^3 + bx^2 + cx + d$.

Degree The highest exponent in a polynomial. In $x^3 + 5x^2 + 7x + 2$ the degree is 3.

Degree A unit of angle measurement that is $\dfrac{1}{360}$ of a circle.

Dependent events When the probability of one event is affected by whether or not some other event happened, they are called dependent events. The event "The baseball game will be canceled" and the event "It will rain" are dependent events.

Difference of perfect squares A way of factoring a polynomial like $x^2 - 3^2$ into $(x - 3)(x + 3)$. In general, an expression of the form $a^2 - b^2$ can be factored into $(a - b)(a + b)$.

Directrix Every point on a parabola is equidistant from a point and a line. That line is called the directrix.

e A mathematical constant that is involved in many questions related to modeling growth. It is approximately 2.72.

Elimination method Multiplying one or more equations in a system of equations by a constant and then adding the equations in a way that makes one of the variables "cancel out." The system $x + 2y = 5$ and $3x - 2y = 7$ becomes $4x = 12$ when the two equations are added together.

Even function A function whose graph is symmetric with respect to the y-axis.

Experimental study When a researcher has one group of subjects do one thing and another group do another thing and studies how the two groups react. An example is giving half of the subjects one type of medicine and the rest of them another type of medicine.

Explicit sequence formula A way of describing a sequence based on relating the position of a term to the number in that position. $a_n = 5n + 8$ is an explicit sequence formula.

Exponential equation An equation that involves an exponential expression. $3 \cdot 2^x = 96$ is an exponential equation.

Exponential expression An expression that has a variable as an exponent. $3 \cdot 2^x$ is an exponential expression.

Exponential function A function that has a variable as an exponent. $f(x) = 3 \cdot 2^x$ is an exponential function.

Factor by grouping A way of factoring that involves separating a polynomial into two or more groups, finding a common factor in each group, and, if possible, factoring a common factor out of what remains. The expression

$$x^3 + 3x^2 + 2x + 6 = x^2(x + 3) + 2(x + 3) = (x + 3)(x^2 + 2)$$

is an example of factoring by grouping.

Factoring Finding two things whose product is a given thing. $15 = 3 \cdot 5$ is a way of factoring 15. $x^2 - 5x + 6 = (x - 2)(x - 3)$ is a way of factoring $x^2 - 5x + 6$.

Factors When a number or polynomial is divided by a number or polynomial and there is no remainder, the divisor is a factor of the dividend. Since $\frac{15}{3} = 5$ remainder 0, 3 is a factor of 15.

Factor theorem A theorem that states that if the remainder when you divide $P(x)$ by $x - a$ is 0, then $x - a$ is a factor of $P(x)$.

For example, if you divide $P(x) = x^2 + 5x + 6$ by $x + 2$, you get $x + 3$ remainder 0 so $x + 2$ is a factor of $P(x)$.

Finite arithmetic series formula A formula $S_n = \dfrac{(a_1 + a_L)n}{2}$ for summing the terms of a finite arithmetic series.

Finite geometric series formula A formula $S_n = \dfrac{a_1(1 - r^n)}{1 - r}$ for summing the terms of a finite geometric series.

Focus Every point on a parabola is equidistant from a point and a line. That point is called the focus of the parabola.

FOIL A way to remember the four multiplications when multiplying one binomial by another binomial. When multiplying $(2x + 3)(4x + 5)$, F (for firsts) represents $2x \cdot 4x$, O (for outers) represents $2x \cdot 5$, I (for inners) represents $3 \cdot 4x$, and L (for lasts) represents $3 \cdot 5$.

Frequency The number of sine or cosine curves that can fit into a 360-degree interval.

Function A rule that takes a number or expression as an input and converts it into a number or expression in a predicable way. The function $f(x) = x^2$ takes a number or expression and turns it into the number or expression squared, for example, $f(3) = 3^2 = 9$.

Geometric sequence A sequence of numbers like 3, 6, 12, 24, . . . , where each number is equal to the previous number in the sequence multiplied by the same number. That number, the common ratio, is called r. In this case, $r = 2$.

Geometric series The sum of the terms in a geometric sequence. $3 + 6 + 12 + 24$ is a geometric series.

Given When calculating the probability of B given A, the "given" means that the event A has already happened, which may or may not affect the probability that B will also happen.

Greatest common factor Of all the common factors between two numbers or polynomials, the one that is the largest. For example, the numbers 12 and 18 have common factors 1, 2, 3, and 6 so 6 is the greatest common factor between 12 and 18.

Growth rate In an exponential expression $a(1 + r)^x$, the r is the growth rate. For example, in $100(1 + 0.07)^x$, the growth rate is 0.07.

Horizontal shift When all the points of a graph move right or left the same number of units.

Horizontal shrink When the x-coordinates of every point in a graph are divided by the same number.

i A shorthand way of writing $\sqrt{-1}$.

Imaginary number A number that is a multiple of the imaginary unit i. $5i$ is an imaginary number.

Independent events When the probability of one event is not affected by whether or not some other event happened. The event "I wake up at 7:00 A.M." and "It will rain today" are independent events.

Inverse function If a function $f(x)$ has an inverse, $f^{-1}(x)$, then $f^{-1}(f(x)) = x$. For example, if $f(x) = x + 1$, $f^{-1}(x) = x - 1$. An inverse function "undoes" whatever the original function did to it.

Like terms When two monomials with the same variable part are added or subtracted. In the sum $2x^2 + 5x + 3x^2 + 7$, the $2x^2$ and the $3x^2$ are like terms and can be combined to $5x^2$.

Linear function A function whose largest exponent is 1. $f(x) = 3x + 7$ is a linear function.

ln x A shorthand for writing $\log_e x$.

$\log_b x$ The number to which b would have to be raised to get a result of x. For example, $\log_5 125 = 3$ since $5^3 = 125$.

Logarithm An exponent that tells what a base needs to be raised to in order to become another number. The logarithm of 25 base 5 or $\log_5 25 = 2$ since $5^2 = 25$.

Monomial A polynomial that has just one term. The polynomial $3x^5$ is a monomial.

Normal distribution When a set of data points resembles a bell curve with 68% being within one standard deviation and 95% being within two standard deviations.

Observational study When a statistical researcher collects data by studying the behavior of the subjects without interfering with their environment.

Odd function A function whose graph is symmetric with respect to the origin.

Perfect square trinomial A polynomial of the form $x^2 + 2ax + a^2$ or $x^2 - 2ax + a^2$ that can be factored into $(x + a)^2$ or $(x - a)^2$. For example, $x^2 + 6x + 9$ can be factored into $(x + 3)^2$.

Period The number of degrees (or radians) in one cycle of a sine or cosine curve.

Polynomial An expression like $2x^3 + 5x^2 - 3x + 7$ that consists of one or more terms, each term having a coefficient and a variable raised to a power.

Population mean The average of an entire population for some characteristic. For example, if out of 10,000 people studied, their average height is 64 inches, 64 is the population mean.

Population proportion The percent of an entire population that has some characteristic. For example, if out of 10,000 people studied, 52% are boys, 0.52 is the population proportion.

Probability The chance of something happening ranging from 0 (impossible) to 1 (certain).

Quadratic equation An equation of the form $ax^2 + bx + c = 0$.

Quadratic formula The formula $\dfrac{-b \pm \sqrt{b^2 - 4ac}}{2a}$ used to find the solutions to the equation $ax^2 + bx + c = 0$.

Quadratic function A function whose largest exponent is 2. $f(x) = x^2 - 3x + 7$ is a quadratic function.

Radian A unit of angle measurement that is equal to approximately 57 degrees.

Radical equation An equation involving a radical expression. $\sqrt{x+2} = 5$ is a radical equation.

Radical expression An expression involving a radical symbol. $\sqrt{x+2}$ is a radical expression.

Randomization test When a computer is used to simulate an ideal experiment with the goal of seeing if something that already happened was likely to happen or not.

Rational equation An equation that involves one or more rational expressions. $\dfrac{3}{x-2} + \dfrac{4}{x-2} = 5$ is a rational equation.

Rational expression An expression that has a fraction with a polynomial of degree at least one in the denominator. $\dfrac{3}{x+1}$ is a rational expression.

Rational function A function that has a fraction with a polynomial of degree at least one in the denominator. $f(x) = \dfrac{3}{x+1}$ is a rational function.

Recursive sequence formula A way of describing a sequence based on how each term relates to the preceding term. $a_1 = 3$, $a_n = 5 + a_{n-1}$ is a recursive sequence formula.

Remainder theorem A theorem that states that when you divide a polynomial $P(x)$ by $x - a$, the remainder will always equal $P(a)$. For example, if you divide the polynomial $P(x) = x^2 + 5x + 6$ by $x - 3$, you will get a remainder of $P(3) = 30$.

Sample mean The average of a sample that has some characteristic. For example, if out of 10,000 people, 100 people are randomly chosen and, of those 100 people, their average height is 61 inches, then 61 is the sample mean for that sample.

Sample proportion The percent of a sample that has some characteristic. For example, if out of 10,000 people, 100 people are randomly chosen and, of those 100 people, 38% are boys, then 0.38 is the sample proportion.

Sample space A list of all the possible outcomes of a probability experiment.

Sampling distribution of the sample mean When multiple groups of samples are taken from a larger population and the average of some attribute of each sample group is calculated and graphed in a histogram.

Sampling distribution of the sample proportion When multiple groups of samples are taken from a larger population and the percent of some attribute of each sample group is calculated and graphed in a histogram.

Simplified radical form When the number inside a square root (or cube root) sign has no factors that are perfect squares (or perfect cubes). The expression $\sqrt{18}$ is not in simplified radical form since $18 = 2 \cdot 9$. $3\sqrt{2}$, which is equal to $\sqrt{18}$, is in simplest radical form.

Sine The ratio between the opposite side and hypotenuse of a right triangle. Also the y-coordinate of a point on the unit circle.

Standard deviation A way of measuring how "spread out" a data set is. The smallest standard deviation is 0 when all the elements of a set are equal.

Standard form of a quadratic expression An expression of the form $y = ax^2 + bx + c$.

Statistical significance When data from an experiment are randomly mixed together multiple times with the goal of seeing if a difference between two groups was likely to happen.

Substitution method In a system of equations, when one of the variables is solved in terms of the others and then substituted into the other equation. For example, the system of equations $y = 2x + 1$, $3x + 2y = 10$, the y in the second equation can be replaced by $2x + 1$ to become $3x + 2(2x + 1) = 10$.

Survey A question or set of questions answered by a subject in a statistical study.

System of equations Two or more equations that have two or more variables. A system of equations generally has one solution, which is an ordered pair (or triple) with numbers that satisfy each of the equations.

Tangent The ratio between the opposite side and the adjacent side of a right triangle.

Trigonometric function A function that involves a trigonometry ratio, like sine, cosine, or tangent. $f(x) = 3 \sin(x) + 1$ is a trigonometric function.

Trigonometric identity Two expressions involving trigonometric ratios that can be shown to be equal to one another.

Trinomial A polynomial with three terms. The expression $x^2 + 5x + 6$ is a trinomial.

Two-way frequency table A chart used for listing the different ways that things can have or not have certain attributes.

Unit circle A circle with a radius of 1 and center at $(0, 0)$.

Variable A letter that is used to represent a number. Often this letter represents the thing to be solved for in an equation. In the expression $2x + 5$, the variable is x.

Venn diagram A way of organizing data that have two or more intersecting circles surrounded by a rectangle.

Vertex form of a quadratic expression An expression of the form $y = a(x - h)^2 + k$. When graphed, it is a parabola with vertex (h, k).

Vertical shift When all the points of a graph move up or down the same number of units.

Vertical stretch When the y-coordinates of every point in a graph are multiplied by the same number.

z-score A measure of how many standard deviations above (positive *z-score*) or below (negative *z-score*) a number is from the mean.

Regents Examinations, Answers, and Self-Analysis Charts

Examination
June 2016
Algebra II

HIGH SCHOOL MATH REFERENCE SHEET

Conversions

1 inch = 2.54 centimeters	1 cup = 8 fluid ounces
1 meter = 39.37 inches	1 pint = 2 cups
1 mile = 5280 feet	1 quart = 2 pints
1 mile = 1760 yards	1 gallon = 4 quarts
1 mile = 1.609 kilometers	1 gallon = 3.785 liters
	1 liter = 0.264 gallon
1 kilometer = 0.62 mile	1 liter = 1000 cubic centimeters
1 pound = 16 ounces	
1 pound = 0.454 kilogram	
1 kilogram = 2.2 pounds	
1 ton = 2000 pounds	

Formulas

Triangle	$A = \dfrac{1}{2}bh$
Parallelogram	$A = bh$
Circle	$A = \pi r^2$
Circle	$C = \pi d$ or $C = 2\pi r$

Formulas (continued)

General Prisms	$V = Bh$
Cylinder	$V = \pi r^2 h$
Sphere	$V = \dfrac{4}{3}\pi r^3$
Cone	$V = \dfrac{1}{3}\pi r^2 h$
Pyramid	$V = \dfrac{1}{3}Bh$
Pythagorean Theorem	$a^2 + b^2 = c^2$
Quadratic Formula	$x = \dfrac{-b \pm \sqrt{b^2 - 4ac}}{2a}$
Arithmetic Sequence	$a_n = a_1 + (n-1)d$
Geometric Sequence	$a_n = a_1 r^{n-1}$
Geometric Series	$S_n = \dfrac{a_1 - a_1 r^n}{1 - r}$ where $r \neq 1$
Radians	1 radian $= \dfrac{180}{\pi}$ degrees
Degrees	1 degree $= \dfrac{\pi}{180}$ radians
Exponential Growth/Decay	$A = A_0 e^{k(t - t_0)} + B_0$

PART I

Answer all 24 questions in this part. Each correct answer will receive 2 credits. No partial credit will be allowed. For each statement or question, write in the space provided the numeral preceding the word or expression that best completes the statement or answers the question.   [48 credits]

1 When $b > 0$ and d is a positive integer, the expression $(3b)^{\frac{2}{d}}$ is equivalent to

(1) $\dfrac{1}{\left(\sqrt[d]{3b}\right)^2}$

(3) $\dfrac{1}{\sqrt{3b^d}}$

(2) $\left(\sqrt{3b}\right)^d$

(4) $\left(\sqrt[d]{3b}\right)^2$

1 _____

2 Julie averaged 85 on the first three tests of the semester in her mathematics class. If she scores 93 on each of the remaining tests, her average will be 90. Which equation could be used to determine how many tests, T, are left in the semester?

(1) $\dfrac{255+93T}{3T} = 90$

(3) $\dfrac{255+93T}{T+3} = 90$

(2) $\dfrac{255+90T}{3T} = 93$

(4) $\dfrac{255+90T}{T+3} = 93$

2 _____

3 Given i is the imaginary unit, $(2 - yi)^2$ in simplest form is

(1) $y^2 - 4yi + 4$

(3) $-y^2 + 4$

(2) $-y^2 - 4yi + 4$

(4) $y^2 + 4$

3 _____

4 Which graph has the following characteristics?

- three real zeros
- as $x \to -\infty$, $f(x) \to -\infty$
- as $x \to \infty$, $f(x) \to \infty$

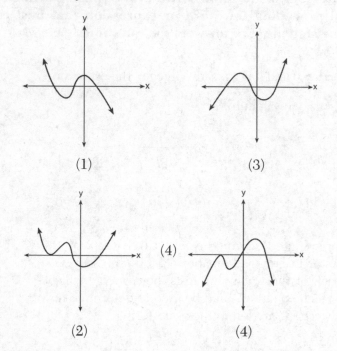

(1)   (3)

(2)   (4)   (4)   4 _____

5 The solution set for the equation is $\sqrt{56 - x} = x$ is

(1) $\{-8, 7\}$ (3) $\{7\}$

(2) $\{-7, 8\}$ (4) $\{\ \}$ 5 _____

6 The zeros for $f(x) = x^4 - 4x^3 - 9x^2 + 36x$ are

(1) $\{0, \pm 3, 4\}$ (3) $\{0, \pm 3, -4\}$

(2) $\{0, 3, 4\}$ (4) $\{0, 3, -4\}$ 6 _____

7 Anne has a coin. She does not know if it is a fair coin. She flipped the coin 100 times and obtained 73 heads and 27 tails. She ran a computer simulation of 200 samples of 100 fair coin flips. The output of the proportion of heads is shown below.

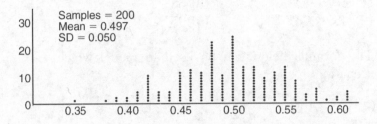

Given the results of her coin flips and of her computer simulation, which statement is most accurate?

(1) 73 of the computer's next 100 coin flips will be heads.

(2) 50 of her next 100 coin flips will be heads.

(3) Her coin is not fair.

(4) Her coin is fair.

7 _____

8 If $g(c) = 1 - c^2$ and $m(c) = c + 1$, then which statement is *not* true?

(1) $g(c) \cdot m(c) = 1 + c - c^2 - c^3$

(2) $g(c) + m(c) = 2 + c - c^2$

(3) $m(c) - g(c) = c + c^2$

(4) $\dfrac{m(c)}{g(c)} = \dfrac{-1}{1-c}$

8 _____

9 The heights of women in the United States are normally distributed with a mean of 64 inches and a standard deviation of 2.75 inches. The percent of women whose heights are between 64 and 69.5 inches, to the *nearest whole percent*, is

(1) 6   (3) 68

(2) 48   (4) 95   9 _____

10 The formula below can be used to model which scenario?

$$a_1 = 3000$$
$$a_n = 0.80a_{n-1}$$

(1) The first row of a stadium has 3000 seats, and each row thereafter has 80 more seats than the row in front of it.

(2) The last row of a stadium has 3000 seats, and each row before it has 80 fewer seats than the row behind it.

(3) A bank account starts with a deposit of $3000, and each year it grows by 80%.

(4) The initial value of a specialty toy is $3000, and its value each of the following years is 20% less.   10 _____

11 Sean's team has a baseball game tomorrow. He pitches 50% of the games. There is a 40% chance of rain during the game tomorrow. If the probability that it rains given that Sean pitches is 40%, it can be concluded that these two events are

(1) independent   (3) mutually exclusive

(2) dependent   (4) complements   11 _____

12 A solution of the equation $2x^2 + 3x + 2 = 0$ is

(1) $-\dfrac{3}{4} + \dfrac{1}{4} i\sqrt{7}$ (3) $-\dfrac{3}{4} + \dfrac{1}{4} \sqrt{7}$

(2) $-\dfrac{3}{4} + \dfrac{7}{4} i$ (4) $\dfrac{1}{2}$ 12 _____

13 The Ferris wheel at the landmark Navy Pier in Chicago takes 7 minutes to make one full rotation. The height, H, in feet, above the ground of one of the six-person cars can be modeled by

$H(t) = 70 \sin\left(\dfrac{2\pi}{7}(t - 1.75)\right) + 80$, where t is time, in minutes. Using $H(t)$ for one full rotation, this car's minimum height, in feet, is

(1) 150   (3) 10
(2) 70   (4) 0   13 _____

14 The expression $\dfrac{4x^3 + 5x + 10}{2x + 3}$ is equivalent to

(1) $2x^2 + 3x - 7 + \dfrac{31}{2x + 3}$ (3) $2x^2 + 2.5x + 5 + \dfrac{15}{2x + 3}$

(2) $2x^2 - 3x + 7 - \dfrac{11}{2x + 3}$ (4) $2x^2 - 2.5x - 5 - \dfrac{20}{2x + 3}$ 14 _____

15 Which function represents exponential decay?

(1) $y = 2^{0.3t}$ (3) $y = \left(\dfrac{1}{2}\right)^{-t}$

(2) $y = 1.2^{3t}$ (4) $y = 5^{-t}$ 15 _____

16 Given $f^{-1}(x) = -\dfrac{3}{4}x + 2$, which equation represents $f(x)$?

(1) $f(x) = \dfrac{4}{3}x - \dfrac{8}{3}$ (3) $f(x) = \dfrac{3}{4}x - 2$

(2) $f(x) = -\dfrac{4}{3}x + \dfrac{8}{3}$ (4) $f(x) = -\dfrac{3}{4}x + 2$ 16 _____

17 A circle centered at the origin has a radius of 10 units. The terminal side of an angle, θ, intercepts the circle in quadrant II at point C. The y-coordinate of point C is 8. What is the value of cos θ?

(1) $-\dfrac{3}{5}$ (3) $\dfrac{3}{5}$

(2) $-\dfrac{3}{4}$ (4) $\dfrac{4}{5}$ 17 _____

18 Which statement about the graph of $c(x) = \log_6 x$ is *false*?

(1) The asymptote has equation $y = 0$.
(2) The graph has no y-intercept.
(3) The domain is the set of positive reals.
(2) The range is the set of all real numbers.   18 _____

19 The equation $4x^2 - 24x + 4y^2 + 72y = 76$ is equivalent to

(1) $4(x - 3)^2 + 4(y + 9)^2 = 76$
(2) $4(x - 3)^2 + 4(y + 9)^2 = 121$
(3) $4(x - 3)^2 + 4(y + 9)^2 = 166$
(2) $4(x - 3)^2 + 4(y + 9)^2 = 436$ 19 _____

20 There was a study done on oxygen consumption of snails as a function of pH, and the result was a degree 4 polynomial function whose graph is shown below.

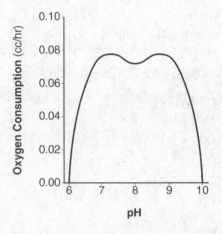

Which statement about this function is *incorrect*?

(1) The degree of the polynomial is even.
(2) There is a positive leading coefficient.
(3) At two pH values, there is a relative maximum value.
(4) There are two intervals where the function is decreasing.   20 _____

21 Last year, the total revenue for Home Style, a national restaurant chain, increased 5.25% over the previous year. If this trend were to continue, which expression could the company's chief financial officer use to approximate their monthly percent increase in revenue? [Let m represent months.]

(1) $(1.0525)^m$ (3) $(1.00427)^m$

(2) $(1.0525)^{\frac{12}{m}}$ (4) $(1.00427)^{\frac{m}{12}}$ 21 _____

22 Which value, to the *nearest tenth*, is *not* a solution of
$p(x) = q(x)$ if $p(x) = x^3 + 3x^2 - 3x - 1$ and $q(x) = 3x + 8$?

(1) -3.9 (3) 2.1

(2) -1.1 (4) 4.7 22 _____

23 The population of Jamesburg for the years
2010–2013, respectively, was reported as follows:

250,000   250,937   251,878   252,822

How can this sequence be recursively modeled?

(1) $j_n = 250,000(1.00375)^{n-1}$

(2) $j_n = 250,000 + 937^{(n-1)}$

(3) $j_1 = 250,000$
    $j_n = 1.00375j_{n-1}$

(4) $j_1 = 250,000$
    $j_n = j_{n-1} + 937$ 23 _____

24 The voltage used by most households can be modeled
by a sine function. The maximum voltage is 120 volts,
and there are 60 cycles *every second*. Which equation best represents the value of the voltage as it flows
through the electric wires, where t is time in seconds?

(1) $V = 120 \sin(t)$ (3) $V = 120 \sin(60\pi t)$

(2) $V = 120 \sin(60t)$ (4) $V = 120 \sin(120\pi t)$ 24 _____

PART II

Answer all 8 questions in this part. Each correct answer will receive 2 credits. Clearly indicate the necessary steps, including appropriate formula substitutions, diagrams, graphs, charts, etc. For all questions in this part, a correct numerical answer with no work shown will receive only 1 credit.   [16 credits]

25   Solve for x: $\dfrac{1}{x} - \dfrac{1}{3} = -\dfrac{1}{3x}$

26 Describe how a controlled experiment can be created to examine the effect of ingredient X in a toothpaste.

27 Determine if $x - 5$ is a factor of $2x^3 - 4x^2 - 7x - 10$. Explain your answer.

28 On the axes below, graph *one* cycle of a cosine function with amplitude 3, period $\frac{\pi}{2}$, midline $y = -1$, and passing through the point $(0, 2)$.

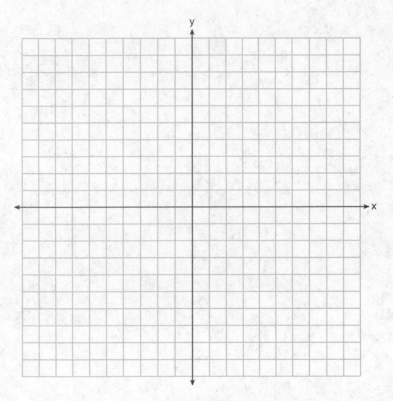

29 A suburban high school has a population of 1376 students. The number of students who participate in sports is 649. The number of students who participate in music is 433. If the probability that a student participates in either sports or music is $\frac{974}{1376}$, what is the probability that a student participates in both sports and music?

30 The directrix of the parabola $12(y + 3) = (x - 4)^2$ has the equation $y = -6$. Find the coordinates of the focus of the parabola.

31 Algebraically prove that $\dfrac{x^3 + 9}{x^3 + 8} = 1 + \dfrac{1}{x^3 + 8}$, where $x \neq -2$.

32 A house purchased 5 years ago for $100,000 was just sold for $135,000. Assuming exponential growth, approximate the annual growth rate, to the *nearest percent*.

PART III

Answer all 4 questions in this part. Each correct answer will receive 4 credits. Clearly indicate the necessary steps, including appropriate formula substitutions, diagrams, graphs, charts, etc. For all questions in this part, a correct numerical answer with no work shown will receive only 1 credit.   [16 credits]

33 Solve the system of equations shown below algebraically.

$$(x - 3)^2 + (y + 2)^2 = 16$$
$$2x + 2y = 10$$

34 Alexa earns $33,000 in her first year of teaching and earns a 4% increase in each successive year.

Write a geometric series formula, S_n, for Alexa's total earnings over n years.

Use this formula to find Alexa's total earnings for her first 15 years of teaching, to the *nearest cent*.

35 Fifty-five students attending the prom were randomly selected to participate in a survey about the music choice at the prom. Sixty percent responded that a DJ would be preferred over a band. Members of the prom committee thought that the vote would have 50% for the DJ and 50% for the band.

A simulation was run 200 times, each of sample size 55, based on the premise that 60% of the students would prefer a DJ. The approximate normal simulation results are shown below.

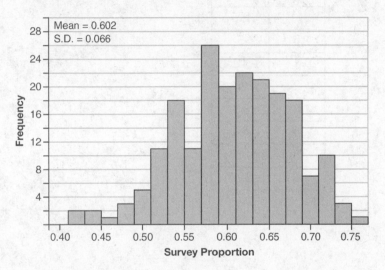

Using the results of the simulation, determine a plausible interval containing the middle 95% of the data. Round all values to the *nearest hundredth*.

Members of the prom committee are concerned that a vote of all students attending the prom may produce a 50%–50% split. Explain what statistical evidence supports this concern.

36 Which function shown below has a greater average rate of change on the interval $[-2, 4]$? Justify your answer.

x	f(x)
−4	0.3125
−3	0.625
−2	1.25
−1	2.5
0	5
1	10
2	20
3	40
4	80
5	160
6	320

$g(x) = 4x^3 - 5x^2 + 3$

PART IV

Answer the question in this part. A correct answer will receive 6 credits. Clearly indicate the necessary steps, including appropriate formula substitutions, diagrams, graphs, charts, etc. A correct numerical answer with no work shown will receive only 1 credit. [6 credits]

37 Drugs break down in the human body at different rates and therefore must be prescribed by doctors carefully to prevent complications, such as overdosing. The breakdown of a drug is represented by the function $N(t) = N_0(e)^{-rt}$, where $N(t)$ is the amount left in the body, N_0 is the initial dosage, r is the decay rate, and t is time in hours. Patient A, $A(t)$, is given 800 milligrams of a drug with a decay rate of 0.347. Patient B, $B(t)$, is given 400 milligrams of another drug with a decay rate of 0.231.

Write two functions, $A(t)$ and $B(t)$, to represent the breakdown of the respective drug given to each patient.

Question 37 is continued on the next page.

Question 37 continued.

Graph each function on the set of axes below.

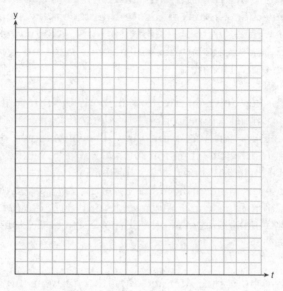

To the *nearest hour*, *t*, when does the amount of the given drug remaining in patient *B* begin to exceed the amount of the given drug remaining in patient *A*?

The doctor will allow patient *A* to take another 800 milligram dose of the drug once only 15% of the original dose is left in the body. Determine, to the *nearest tenth of an hour*, how long patient *A* will have to wait to take another 800 milligram dose of the drug.

Answers
June 2016
Algebra II

Answer Key

PART I

1. (4)	**5.** (3)	**9.** (2)	**13.** (3)	**17.** (1)	**21.** (3)
2. (3)	**6.** (1)	**10.** (4)	**14.** (2)	**18.** (1)	**22.** (4)
3. (2)	**7.** (3)	**11.** (1)	**15.** (4)	**19.** (4)	**23.** (3)
4. (3)	**8.** (4)	**12.** (1)	**16.** (2)	**20.** (2)	**24.** (4)

PART II

25. 4

26. The experimenter must randomly split subjects into a control group and an experimental group.

27. No

28.

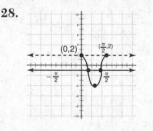

29. $\dfrac{108}{1376}$

30. $(4, 0)$

31. Yes

32. 6%

PART III

33. $(3, 2)$ and $(7, -2)$

34. $S_n = \dfrac{33,000 - 33,000 \cdot 1.04^n}{1 - 1.04}$,

660,778.39

35. Between 0.47 and 0.73

36. g

PART IV

37. $A(t) = 800(e)^{-0.347t}$,
$B(t) = 400(e)^{-0.231t}$, 6, 5.5

In **Parts II–IV**, you are required to show how you arrived at your answers. For sample methods of solutions, see the *Answers Explained* section.

Answers Explained

PART I

1. An expression of the form $x^{\frac{a}{b}}$ involving a fractional exponent is equivalent to the expression $\left(\sqrt[b]{x}\right)^{a}$. So $(3b)^{\frac{2}{d}} = \left(\sqrt[d]{3b}\right)^{2}$.

   The correct choice is (**4**).

2. Use the formula that relates the average, the total points, and the number of tests:

   $$\frac{\text{total points}}{\text{number of tests}} = \text{average}$$

   Use this formula to determine how many total points Julie had after taking the first three tests:

   $$\frac{\text{total points}}{3} = 85$$

   $$3 \times \frac{\text{total points}}{3} = 3 \times 85$$

   $$= 255$$

   If Julie gets a 93 on T more tests, her total points will increase by $93T$ and the number of tests will increase by T. If the new average is known to be 90, the equation relating the new average, the new total points, and the new number of tests can be determined:

   $$\frac{255 + 93T}{3 + T} = 90$$

   The correct choice is (**3**).

3. Expand $(2 - yi)^2$ as $(2 - yi)(2 - yi)$ and multiply with FOIL:

   $$(2 - yi)(2 - yi) = 4 - 2yi - 2yi + y^2 i^2$$

   Since $i^2 = -1$, this is equivalent to:

   $$4 - 2yi - 2yi - y^2 = 4 - 4yi - y^2$$
   $$= -y^2 - 4yi + 4$$

   The correct choice is (**2**).

4. Real zeros of a function correspond to x-intercepts of the graph of the function. Each of the four choices have three x-intercepts, so they each have three real zeros. For the other two characteristics, the graph must decrease as x gets more and more negative and must increase as x gets more and more positive.

The correct choice is **(3)**.

5. Since the radical is already isolated, square both sides and solve the resulting quadratic equation:

$$\sqrt{56-x} = x^2$$
$$\left(\sqrt{56-x}\right)^2 = x^2$$
$$56-x = x^2$$
$$0 = x^2 + x - 56$$
$$0 = (x+8)(x-7)$$
$$x+8=0 \quad \text{or} \quad x-7=0$$
$$x=-8 \text{ or} \qquad x=7$$

When squaring both sides of a radical equation, there is a chance that an extraneous solution is introduced. So you must check both answer choices:

Check $x = 7$:

$$\sqrt{56-7} = 7$$
$$\sqrt{49} = 7$$
$$7 = 7 \quad ✔$$

Check $x = -8$:

$$\sqrt{56-(-8)} = -8$$
$$\sqrt{64} = -8$$
$$8 = -8 \quad \text{NO}$$

The -8 must be rejected.

The correct choice is **(3)**.

6. The zeros of a function f are values of x that make $f(x) = 0$. The zeros of this function are the solutions to the equation $0 = x^4 - 4x^3 - 9x^2 + 36x$. There are three ways to find the zeros of this function.

Method 1: Factoring

$0 = x^4 - 4x^3 - 9x^2 + 36x$

$0 = x(x^3 - 4x^2 - 9x + 36)$ Factor out the greatest common factor

$0 = x(x^2(x - 4) - 9(x - 4))$ Factor by grouping

$0 = (x - 4)(x^2 - 9)$

$0 = x(x - 4)(x - 3)(x + 3)$ Difference of perfect square factoring

$x = 0$ or $x - 4 = 0$ or $x - 3 = 0$ or $x + 3 = 0$

$x = 0$ or $x = 4$ or $x = 3$ or $x = -3$

Method 2: Graphing Calculator

Find the x-intercepts of the graph of $y = x^4 - 4x^3 - 9x^2 + 36x$. If the x-intercepts are integers, as they are for this graph, they can be easily seen on the graph.

For the TI-84:

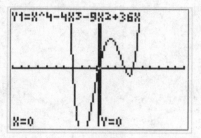

For the TI-Nspire:

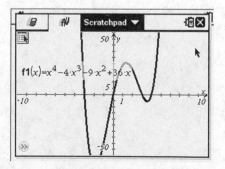

Method 3: Test the Answer Choices

Find the values of $f(0), f(-3), f(3), f(-4)$, and $f(4)$ to see which of them equal 0.

$$f(0) = (0)^4 - 4(0)^3 - 9(0)^2 + 36(0)$$
$$= 0 \quad \checkmark$$

$$f(-3) = (-3)^4 - 4(-3)^3 - 9(-3)^2 + 36(-3)$$
$$= 0 \quad \checkmark$$

$$f(3) = (3)^4 - 4(3)^3 - 9(3)^2 + 36(3)$$
$$= 0 \quad \checkmark$$

$$f(-4) = (-4)^4 - 4(-4)^3 - 9(-4)^2 + 36(-4)$$
$$= 224 \quad \textbf{NO}$$

$$f(4) = (4)^4 - 4(4)^3 - 9(4)^2 + 36(4)$$
$$= 0 \quad \checkmark$$

The correct choice is **(1)**.

7. In 200 simulations there were never any cases where more than 65 heads happened. This suggests that it is very unlikely for a fair coin to land on heads 73 out of 100 flips.

The numbers from the simulations indicate that it is very unlikely for an event to happen that is more than 2 standard deviations (SD) from the mean or less than 2 standard deviations from the mean. In this case, mean + 2SD = 0.597 and mean − 2SD = 0.397. This means that it is very unlikely for there to be more than 60 heads or less than 40 heads if a fair coin is tossed 100 times. Since 73 is more than 60 heads, it is very likely that the coin Anne has is not fair.

The correct choice is **(3)**.

8. Test the four answer choices:

$$g(c) \cdot m(c) = (1-c^2) \cdot (c+1)$$
$$= c+1-c^3-c^2$$
$$= 1+c-c^2-c^3 \qquad ✔$$
$$g(c) + m(c) = (1-c^2) + (c+1)$$
$$= 1-c^2+c+1$$
$$= 2+c-c^2 \qquad ✔$$
$$m(c) - g(c) = (c+1) - (1-c^2)$$
$$= c+1-1+c^2$$
$$= c+c^2 \qquad ✔$$
$$\frac{m(c)}{g(c)} = \frac{c+1}{1-c^2}$$
$$= \frac{c+1}{(1-c)(1+c)}$$
$$= \frac{(1+c)}{(1-c)(1+c)}$$
$$= \frac{1}{1-c}$$
$$\neq \frac{-1}{1-c} \qquad \text{NO}$$

The correct choice is **(4)**.

9. The normal cdf function of the graphing calculator is used for this type of question.

For the TI-84:

Press [2nd] [VARS] [2] to get the normalcdf menu. In the "lower" field, put 64. In the "upper" field, put 69.5. In the "μ" field put the mean, which is 64. In the "σ" field, put the standard deviation, which is 2.75. Select "Paste." Press [ENTER].

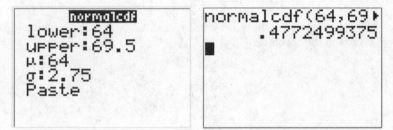

For the TI-Nspire:

Press [A] to open the calculator scratchpad. Press [menu] [6] [5] [2] for the normal cdf menu. In the "Lower Bound" field, put 64. In the "Upper Bound" field put 69.5. In the "μ" field, put the mean, which is 64. In the "σ" field, put the standard deviation, which is 2.75. Select the OK button, and press [enter].

Since 0.47725 rounds to 0.48, 48% of the women have heights between 64 and 69.5 inches.

The correct choice is (**2**).

10. The first number in the sequence is given as $a_1 = 3{,}000$. The second term of the sequence, a_2, can be calculated by substituting 2 for n in the second equation:

$$a_2 = 0.80a_{2-1}$$
$$= 0.80a_1$$
$$= 0.80 \cdot 3{,}000$$
$$= 2{,}400$$

The third term of the sequence, a_3, can be calculated by substituting 3 for n in the second equation:

$$a_3 = 0.80a_{3-1}$$
$$= 0.80a_2$$
$$= 0.80 \cdot 2{,}400$$
$$= 1{,}920$$

The first five terms of this sequence are 3,000; 2,400; 1,920; 1,536; and 1,228.8. Compare this sequence to the four choices:

Choice (1): 3,000; 3,080; 3,160; 3,240; 3,320

Choice (2): 3,000; 2,920; 2,840; 2,760; 2,680

Choice (3): 3,000; 5,400; 9,720; 17,496; 31,492.80

Choice (4): 3,000; 2,400; 1,920; 1,536; 1,228.80

The correct choice is **(4)**.

11. Two events A and B are considered independent if $P(A$ given $B) = P(A)$. If A is the event that it rains and if B is the event that Sean pitches, based on the given information $P(A) = 0.40$ and $P(A$ given $B) = 0.40$. Since these two probabilities are equal, the events are independent.

The correct choice is **(1)**.

12. Use the quadratic formula with $a = 2$, $b = 3$, and $c = 2$:

$$x = \frac{-b \pm \sqrt{b^2 - 4ac}}{2a}$$

$$= \frac{-3 \pm \sqrt{3^2 - 4 \cdot 2 \cdot 2}}{2 \cdot 2}$$

$$= \frac{-3 \pm \sqrt{-7}}{4}$$

$$= \frac{-3 \pm i\sqrt{7}}{4}$$

$$= -\frac{3}{4} \pm \frac{1}{4} i\sqrt{7}$$

The correct choice is **(1)**.

13. Since the minimum value of the sine function is –1, the lowest value of the H function is $H(t) = 70(-1) + 80 = 10$.

Another way to solve this is to graph the function with the graphing calculator and locate the lowest point of the graph. Make sure the calculator is in radian mode.

For the TI-84:

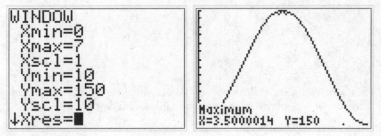

For the TI-Nspire:

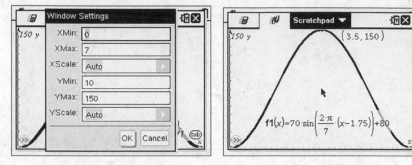

The correct choice is **(3)**.

14. Use polynomial long division:

$$
\begin{array}{r}
2x^2 - 3x + 7 \\
2x+3{\overline{\smash{\big)}\,4x^3 + 0x^2 + 5x + 10}} \\
\underline{-(4x^3 + 6x^2)} \\
-6x^2 + 5x \\
\underline{-(-6x^2 - 9x)} \\
14x + 10 \\
\underline{-(14x + 21)} \\
-11
\end{array}
$$

This remainder, −11, becomes the numerator of the last term. The divisor, $2x + 3$, becomes the denominator of the last term.

The correct choice is **(2)**.

15. Graph each of the four choices on the graphing calculator to see which graph is always decreasing. Of the four choices, only $y = 5^{-t}$ is always decreasing.

For the TI-84:

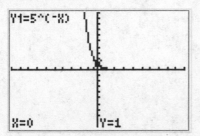

For the TI-Nspire:

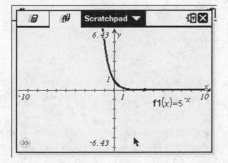

Another way to do this problem is to know that a function with exponential decay is often in the form $y = a^x$ where $0 < a < 1$. Using the properties of exponents, $y = 5^{-t}$ can be expressed as:

$$y = 5^{-1 \cdot t} = \left(5^{-1}\right)^t = \left(\frac{1}{5}\right)^t$$

This has the form of exponential decay.

The correct choice is **(4)**.

16. The inverse of the function $f^{-1}(x)$ is $f(x)$. To find the inverse of a function, follow these three steps:

Step 1: Write the original equation as $y = -\dfrac{3}{4}x + 2$.

Step 2: Swap the x- and the y-variables.

$$x = -\frac{3}{4}y + 2$$

Step 3: Isolate the y in the new equation to find the inverse.

$$x = -\frac{3}{4}y + 2$$

$$\frac{-2 = \qquad -2}{x - 2 = -\frac{3}{4}y}$$

$$-\frac{4}{3}(x - 2) = -\frac{4}{3}\left(-\frac{3}{4}y\right)$$

$$-\frac{4}{3}x + \frac{8}{3} = y$$

This question can also be answered with the graphing calculator. A graph of an inverse function is the graph of the original function reflected over the line $y = x$.

Graph the given inverse function on the same set of axes as each of the choices to see which answer is the reflection of the original equation over the $y = x$-line. The default window for the TI-84 will give a distorted graph, so do [ZOOM] [6] [ZOOM] [5] when graphing.

For the TI-84:   For the TI-Nspire:

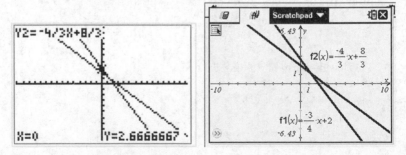

The correct choice is (**2**).

17.

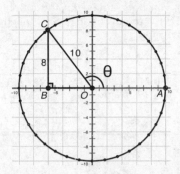

In the diagram, $\angle\theta$ is $\angle AOC$. Triangle OBC is a right triangle with right angle OBC, with $BC = 8$, and with $OC = 10$. Use the sine ratio:

$$\sin\angle BOC = \frac{\text{opposite}}{\text{hypotenuse}} = \frac{8}{10}$$

$$m\angle BOC = \sin^{-1}\left(\frac{8}{10}\right) \approx 53.13°$$

$\angle\theta$ is supplementary to $\angle BOC$, so $m\angle\theta \approx 180° - 53.13° = 126.87°$.

The question asks for the value of $\cos\theta$, and $\cos 126.87° \approx -0.6 = -\dfrac{3}{5}$.

The correct choice is **(1)**.

18. The graph of $c(x)$ can be graphed on the graphing calculator. For the TI-84, use [MATH] [A] to get the logBASE function. For the TI-Nspire, press [ctrl] [10^x].

For the TI-84:

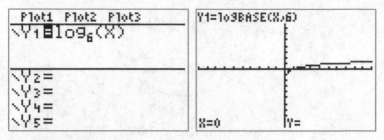

For the TI-Nspire:

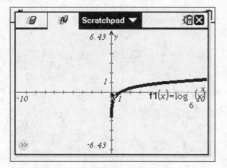

Based on the graph, there is a vertical asymptote at $x = 0$. The y-intercept is $(0, 1)$. The domain is $x > 0$, which is the set of positive reals. The range is the set of all real numbers. There is no horizontal asymptote.

The correct choice is **(1)**.

19. This equation can be converted to an equivalent equation that looks like the answer choices with a technique called "completing the square."

Step 1: Group the x-terms together, and factor out the common factor, 4.

$$4(x^2 - 6x) + 4y^2 + 72y = 76$$

Step 2: Group the y-terms together, and factor out the common factor, 4.

$$4(x^2 - 6x) + 4(y^2 + 18y) = 76$$

Step 3: A trinomial of the form $x^2 + bx + c$ is a perfect square trinomial if $c = \left(\dfrac{b}{2}\right)^2$. In order to make each of the expressions inside the parentheses into perfect square trinomials, the proper value of c needs to be added inside each set of parentheses. For the first set of parentheses, +9 must be added because $\left(\dfrac{-6}{2}\right)^2 = 9$. For the second set of parentheses, +81 must be added because $\left(\dfrac{18}{2}\right)^2 = 81$. On the right side of the equal sign, you must add $4 \times 9 = 36$ and $4 \times 81 = 324$ to keep both sides equal because of the 4s outside the parentheses.

$$4(x^2 - 6x + 9) + 4(y^2 + 18y + 81) = 76 + 36 + 324$$

Step 4: Factor the expressions inside the parentheses, and simplify.

$$4(x - 3)^2 + 4(y + 9)^2 = 436$$

A way to do this question without using completing the square is to simplify each of the answer choices to see which of them becomes the given equation.

The correct choice is **(4)**.

20. Check each of the answer choices.

Choice (1): When a polynomial has an even degree, like 2 or 4, the end behavior at both ends of the graph will be the same. In this graph, the branches at the right and at the left are both decreasing, which suggests that the degree is even. When a polynomial has an odd degree, the end behavior on the two ends will be opposite. One will be increasing, and the other will be decreasing.

Choice (2): When a polynomial has a positive leading coefficient, the end behavior on the right will be increasing. In this graph, the end behavior on the right is decreasing. So the function does not have a positive leading coefficient.

Choice (3): A relative maximum value happens when there is a hump with a highest point on it. In this graph, there is a relative maximum around pH = 7 and another around pH = 9. So there are two pH values where there is a relative maximum value.

Choice (4): Going from left to right, the graph of a decreasing function will be going down. Like a roller coaster that starts as the leftmost point, it would be climbing until it hits pH = 7 and then go down, decreasing, until it reaches pH = 8. Then it would climb again until it reaches pH = 9 and then be decreasing again from pH = 9 and greater. There are two intervals where the function is decreasing.

The correct choice is **(2)**.

21. Since the revenue increases by 5.25% each year, the yearly growth factor is $1 + 0.0525 = 1.0525$. An equation that could be used to calculate the growth factor for y years is $(1.0525)^y$. This could be used to find the percent increase for any number of years. For example, if $y = 10$, the growth factor would be $(1.0525)^{10} = 1.668 = 1 + 0.668$, which would be a percent increase of 66.8%.

To find an equation that involves m for months, replace y with $\frac{m}{12}$, $(1.0525)^{\frac{m}{12}}$. This can be simplified using the rules of exponents.

$$(1.0525)^{\frac{m}{12}} = (1.0525)^{\frac{1}{12} \cdot m}$$

$$= \left((1.0525)^{\frac{1}{12}}\right)^m$$

$$= (1.00427)^m$$

The correct choice is **(3)**.

22. The x-coordinates of the intersection points of the graphs of $p(x)$ and $q(x)$ are the solutions to the equation $p(x) = q(x)$. Using the graphing calculator, graph both functions on the same set of axes. Then find the intersection points. There are three intersection points: -3.9, -1.1, and 2.1.

For the TI-84:

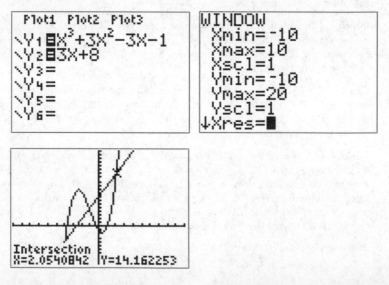

For the TI-Nspire:

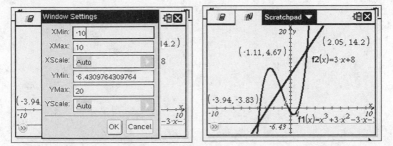

The correct choice is (**4**).

23. Recursive equations have a base case and a recursive part that uses the results of a previous calculation. Choices (**1**) and (**2**) are not recursive equations.

Test choices (**3**) and (**4**) to see which will generate the given numbers.

Choice (3):

$$j_1 = 250,000$$
$$j_2 = 1.00375\,j_{2-1}$$
$$= 1.00375\,j_1$$
$$= 1.00375 \cdot 250,000$$
$$= 250,937.5$$
$$j_3 = 1.00375\,j_{3-1}$$
$$= 1.00375\,j_2$$
$$= 1.00375 \cdot 250,937.5$$
$$= 251,878.5$$
$$j_4 = 1.00375\,j_{4-1}$$
$$= 1.00375\,j_3$$
$$= 1.00375 \cdot 251,878.5$$
$$= 252,823.0$$

Choice (4):

$$j_1 = 250{,}000$$

$$j_2 = j_{2-1} + 937$$
$$= j_1 + 937$$
$$= 250{,}000 + 937$$
$$= 250{,}937$$

$$j_3 = j_{3-1} + 937$$
$$= j_2 + 937$$
$$= 250{,}937 + 937$$
$$= 251{,}874$$

$$j_4 = j_{4-1} + 937$$
$$= j_3 + 937$$
$$= 251{,}874 + 937$$
$$= 252{,}811$$

The numbers generated by choice (3) are closer to the given numbers than the numbers generated by choice (4).

The correct choice is (3).

24. In the equation $V = A \sin(B_t)$, A represents the amplitude. The amplitude is 120 in this scenario since the maximum value is 120. To calculate the B-value, use the equation period $= \dfrac{2\pi}{B}$ when using radians and period $= \dfrac{360°}{B}$ when using degrees. As there is no mention of degrees, it is implied that this equation is intended to be in radians. The period is $\dfrac{1}{60}$ because there are 60 cycles in one second.

Use the formula, $\dfrac{1}{60} = \dfrac{2\pi}{B}$. Solve for B by cross multiplying:

$$B = 60 \cdot 2\pi = 120\pi$$

The correct choice is (4).

PART II

25. Multiply both sides of the equation by the lowest common multiple (LCM) of x, 3, and $3x$, which is $3x$. Then solve for x.

$$3x \cdot \frac{1}{x} - 3x \cdot \frac{1}{3} = 3x \cdot -\frac{1}{3x}$$
$$3 - x = -1$$
$$\underline{-3 = -3}$$
$$-x = -4$$
$$x = 4$$

In a rational equation, you should check your solution(s) by substituting the solution(s) into the original equation to see if any don't make the original equation true.

$$\frac{1}{4} - \frac{1}{3} = -\frac{1}{3 \cdot 4}$$
$$\frac{3}{12} - \frac{4}{12} = -\frac{1}{12}$$
$$-\frac{1}{12} = -\frac{1}{12}$$

The correct answer is $x = 4$.

26. A controlled experiment is when the experimenter makes two groups of subjects. One group is the control group, and the other is the experimental group. Members of the experimental group in this scenario would brush their teeth only with a toothpaste that contains ingredient X. Members of the control group would brush their teeth only with a toothpaste that is the same as the other toothpaste except that it does not contain ingredient X. After some time, data from both groups would be collected and analyzed.

A controlled experiment differs from an observational experiment where the researcher does not control who is using which toothpaste.

27. There are two ways to determine if $x - 5$ is a factor. The long way is to divide $2x^3 - 4x^2 - 7x - 10$ by $x - 5$ and see if there is no remainder.

$$
\begin{array}{r}
2x^2 + 6x + 23 \\
x - 5 \overline{\smash{\big)}\, 2x^3 - 4x^2 - 7x - 10} \\
-\underline{(2x^3 - 10x^2)} \\
6x^2 - 7x \\
-\underline{(6x^2 - 30x)} \\
23x - 10 \\
-\underline{(23x - 115)} \\
105
\end{array}
$$

Since the remainder is 105 and not zero, $x - 5$ is not a factor.

The shortcut for this question is to use the factor theorem. It says that if $x - a$ is a factor of $P(x)$, then $P(a) = 0$:

$$P(5) = 2 \cdot 5^3 - 4 \cdot 5^2 - 7 \cdot 5 - 10$$

$$= 250 - 100 - 35 - 10$$

$$= 105$$

Since $105 \neq 0$, $x - 5$ is not a factor.

28. The graph of a cosine function with midline $y = -1$ and amplitude 3 stays between the lines $y = -1 + 3 = 2$ and $y = -1 - 3 = -4$.

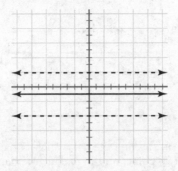

The curve passes through $(0, 2)$, which is the highest point on the graph. So $(0, 2)$ can be the starting point on the graph.

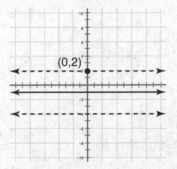

Since the frequency is $\frac{\pi}{2}$, the ending point of the cycle is $\frac{\pi}{2}$ units to the right of the starting point $(0, 2)$, which is $\left(\frac{\pi}{2}, 2\right)$. Since $\frac{\pi}{2}$ is only about 1.5, mark the horizontal scale so that $\frac{\pi}{2}$ is equal to about 5 boxes.

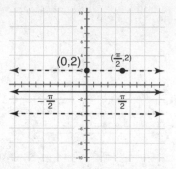

The low point is on the line $y = -4$, which is halfway between the two high points. The two points on the midline are halfway between the high and low points. Connect the 5 points to form a cosine curve.

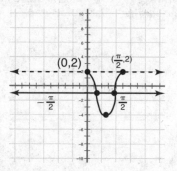

29. The number of students who participate in sports is 649, and the number who participate in music is 433. Together, this is $649 + 433 = 1082$ students. According to the given information, the probability that a student participates in either sports or music is $\frac{974}{1376}$. This means that 974 students participate in sports or in music (or in both sports and music). This means that in the 1082 students calculated by adding the number of students who participate in sports to the number of students who participate in music, some students have been counted twice. These are the students who participate in both sports and music. Subtract 974 from 1082 to get the number of students who have been counted twice: $1082 - 974 = 108$. Since 108 students out of 1376 participate in both sports and music, the probability that a student participates in both is $\frac{108}{1376}$.

This information can be organized on a Venn diagram. The two circles divide the rectangle into four regions. A is the region of students who participate in sports but not in music. C is the region of students who participate in music but not in sports. B is the region of students who participate in both sports and music. D is the region of students who participate in neither sports nor music.

- Regions A and B combined equal 649 students participating in sports.

- Regions B and C combined equal 433 students participating in music.

- Regions A, B, and C combined equal 974 students.

Adding 649 and 433 is the sum of regions A, B, another region B, and region C. This is the sum of 649 and 433, which is 1082. Region B, then, is 1082 (which has two region Bs in it) minus 974. This is given as the sum of regions A, B, and C.

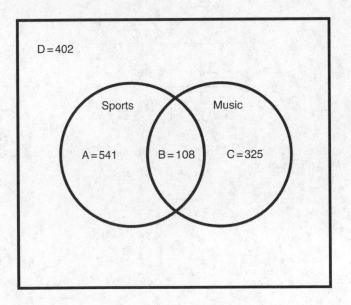

30. When the equation of a parabola is of the form $4p(y - k) = (x - h)^2$, the vertex is (h, k). The graph of this equation is a parabola with a vertex at $(4, -3)$.

    The focus of the parabola is p units above the vertex (or below if p is negative). Since $12 = 4p, p = 3$. So the vertex is 3 units above $(4, -3)$ at $(4, 0)$.

    Even without knowing this formula, this question can be answered another way. Since the directrix of this parabola is 3 units away from the vertex of the parabola, the focus of the parabola will also be 3 units away from the vertex of the parabola but in the opposite direction. The directrix is 3 units below the vertex, so the focus will be 3 units above the vertex, which is located at $(4, 0)$.

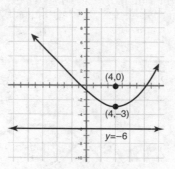

31. Simplify the right side of the equation to show that it is equal to the left side of the equation:

$$\frac{x^3 + 9}{x^3 + 8} = 1 + \frac{1}{x^3 + 8}$$

$$= \frac{x^3 + 8}{x^3 + 8} + \frac{1}{x^3 + 8}$$

$$= \frac{x^3 + 8 + 1}{x^3 + 8}$$

$$= \frac{x^3 + 9}{x^3 + 8}$$

32. The formula new value = old value$(1 + r)^t$ can be used where t is the time in years and r is the annual growth rate:

$$135,000 = 100,000(1+r)^5$$

$$\frac{135,000}{100,000} = \frac{100,000(1+r)^5}{100,000}$$

$$1.35 = (1+r)^5$$

$$\sqrt[5]{1.35} = \sqrt[5]{(1+r)^5}$$

$$1.06 = 1+r$$

$$\underline{-1 = -1}$$

$$0.06 = r$$

The annual growth rate is 6%.

PART III

33. To use the substitution method of solving a system of equations, one of the equations must be converted into an equivalent form in which one of the variables is isolated. The second equation can be quickly solved for x in terms of y:

$$2x + 2y = 10$$
$$\underline{-2y = -2y}$$
$$2x = -2y + 10$$
$$\frac{2x}{2} = \frac{-2y + 10}{2}$$
$$x = -y + 5$$

Replace the x in the first equation with $-y + 10$, and solve for y:

$$(-y + 5 - 3)^2 + (y + 2)^2 = 16$$
$$(-y + 2)^2 + (y + 2)^2 = 16$$
$$y^2 - 4y + 4 + y^2 + 4y + 4 = 16$$
$$2y^2 + 8 = 16$$
$$\underline{-8 = -8}$$
$$2y^2 = 8$$
$$\frac{2y^2}{2} = \frac{8}{2}$$
$$y^2 = 4$$
$$y = \pm 2$$

For each y-value, substitute into either original equation and solve for the corresponding x-value.

For $y = 2$:

$$2x + 2(2) = 10$$
$$2x + 4 = 10$$
$$\underline{-4 = -4}$$
$$2x = 6$$
$$\frac{2x}{2} = \frac{6}{2}$$
$$x = 3$$

For $y = -2$:

$$2x + 2(-2) = 10$$
$$2x - 4 = 10$$
$$+4 = +4$$
$$2x = 14$$
$$\frac{2x}{2} = \frac{14}{2}$$
$$x = 7$$

The solutions are $(3, 2)$ and $(7, -2)$.

34. If Alexa earns \$33,000 the first year, in the second year her salary will increase by $0.04 \cdot \$33,000 = \$1320$. So her salary in the second year will be $\$33,000 + \$1320 = \$34,320$. A quicker way to get this value would be to multiply \$33,000 by 1.04:

$$\$33,000 \cdot 1.04 = \$34,320$$

To get to the salary for the third year, multiply again by 1.04:

$$\$34,320 \cdot 1.04 = \$35,692.80$$

To get this number without first calculating the salary for the second year, you could calculate $\$33,000 \cdot 1.04^2$.

In general, Alexa's salary in year n will be $33,000 \cdot 1.04^{n-1}$. To find her total earnings after n years, you have to use the formula to calculate the sum of the geometric series:

$$S_n = 33,000 + 33,000 \cdot 1.04^1 + 33,000 \cdot 1.04^2 + 33,000 \cdot 1.04^3 + \cdots + 33,000 \cdot 1.04^{n-1}$$

In the reference sheet, the formula for the sum of a geometric series

$S_n = \dfrac{a_1 - a_1 r^n}{1 - r}$ is given. For this example, $a_1 = 33,000$ and $r = 1.04$.

So the formula becomes:

$$S_n = \frac{33,000 - 33,000 \cdot 1.04^n}{1 - 1.04}$$

For $n = 15$:

$$S_{15} = \frac{33,000 - 33,000 \cdot 1.04^{15}}{1 - 1.04} = 660,778.39$$

Alexa's total earnings for her first 15 years of teaching is \$660,778.39.

35. The plausible interval containing the middle 95% of the data is between mean − 2 · SD and mean + 2 · SD. For this example, that is between $0.602 - 2 \cdot 0.066 = 0.47$ and $0.602 + 2 \cdot 0.066 = 0.73$.

The members of the prom committee are correct to be concerned since 0.50 is between 0.47 and 0.73. Alternately, it can be seen on the histogram that 0.50 is something that will happen sometimes, even when 60% of the total population prefers a DJ. If the histogram showed no values equal to or less than 0.50, it would suggest that a 50%–50% was very unlikely.

36. The average rate of change of a function f on an interval $[a, b]$ can be calculated with the formula average rate of change $= \dfrac{f(b) - f(a)}{b - a}$.

The average rate of change for the given function f on the interval $[-2, 4]$ is:

$$\frac{f(4) - f(-2)}{4 - (-2)} = \frac{80 - 1.25}{6}$$
$$= \frac{78.75}{6}$$
$$= 13.125$$

The average rate of change for the given function g on the interval $[-2, 4]$ is:

$$\frac{g(4) - g(-2)}{4 - (-2)} = \frac{[4(4)^3 - 5(4)^2 + 3] - [4(-2)^3 - 5(-2)^2 + 3]}{6}$$
$$= \frac{179 - (-49)}{6}$$
$$= \frac{228}{6}$$
$$= 38$$

Since $38 > 13.125$, the average rate of change on the interval $[-2, 4]$ is greater for g than it is for f.

PART IV

37. For patient A, $N_0 = 800$ and $r = 0.347$. So the equation is $A(t) = 800(e)^{-0.347t}$.

    For patient B, $N_0 = 400$ and $r = 0.231$. So the equation is $B(t) = 400(e)^{-0.231t}$.

    When the graph of both functions is graphed on a graphing calculator and the intersect feature is used, the intersection point is approximately (6, 100). Before hour 6, patient A has more of the drug in his or her body. After hour 6, patient B has more of the drug in his or her body.

    For the TI-84:

    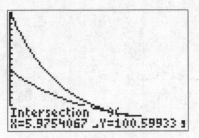

    For the TI-Nspire:

    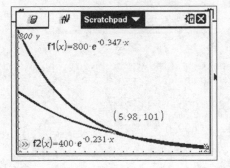

Make your scale so the graph includes the two y-intercepts and also the intersection point. On the axes below, each vertical unit is 40 milligrams and each horizontal unit is $\frac{1}{2}$ hour.

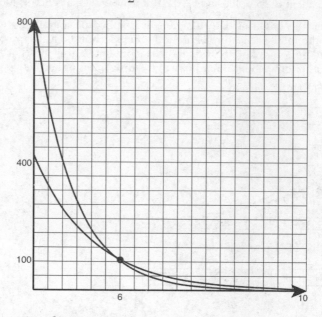

Since 15% of 800 is 120, the next dose will be given when $A(t)=120$. This becomes an exponential equation:

$$120 = 800(e)^{-0.347t}$$

$$\frac{120}{800} = \frac{800(e)^{-0.347t}}{800}$$

$$0.15 = e^{-0.347t}$$

$$\ln 0.15 = \ln e^{-0.347t}$$

$$\ln 0.15 = -0.347t$$

$$\frac{\ln 0.15}{-0.347} = \frac{-0.347t}{-0.347}$$

$$5.5 \approx t$$

The next dose can be administered after 5.5 hours.

This can also be solved with the graphing calculator by finding the x-coordinate of the intersection of $A(x)$ and $y = 120$.

For the TI-84:

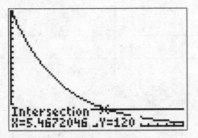

For the TI-Nspire:

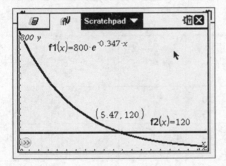

Topic	Question Numbers	Number of Points	Your Points	Your Percentage
1. Polynomial Expressions and Equations	6, 12, 22, 27, 30, 36	2 + 2 + 2 + 2 + 2 + 4 = 14		
2. Complex Numbers	3	2		
3. Exponential Expressions and Equations	1, 15, 21, 32, 34, 37	2 + 2 + 2 + 4 + 4 + 4 = 18		
4. Rational Expressions and Equations	14, 25, 31	2 + 2 + 2 = 6		
5. Radical Expressions and Equations	5	2		
6. Trigonometric Expressions and Equations	13, 17, 24, 28	2 + 2 + 2 + 2 = 8		
7. Graphing	4, 18, 20	2 + 2 + 2 = 6		
8. Functions	8, 16, 19	2 + 2 + 2 = 6		
9. Systems of Equations	33	4		
10. Sequences and Series	10, 23	2 + 2 = 4		
11. Probability	11, 19	2 + 2 = 4		
12. Statistics	2, 7, 9, 26, 35	2 + 2 + 2 + 2 + 4 = 12		

HOW TO CONVERT YOUR RAW SCORE TO YOUR ALGEBRA II REGENTS EXAMINATION SCORE

The accompanying conversion chart must be used to determine your final score on the June 2016 Regents Examination in Algebra II. To find your final exam score, locate in the column labeled "Raw Score" the total number of points you scored out of a possible 86 points. Since partial credit is allowed in Parts II, III, and IV of the test, you may need to approximate the credit you would receive for a solution that is not completely correct. Then locate in the adjacent column to the right the scale score that corresponds to your raw score. The scale score is your final Algebra II Regents Examination score.

Regents Examination in Algebra II—June 2016
Chart for Converting Total Test Raw Scores to Final
Examination Scores (Scaled Scores)

Raw Score	Scale Score	Performance Level	Raw Score	Scale Score	Performance Level	Raw Score	Scale Score	Performance Level
86	100	5	57	82	4	28	68	3
85	99	5	56	82	4	27	67	3
84	98	5	55	81	4	26	66	3
83	97	5	54	81	4	25	65	3
82	97	5	53	81	4	24	64	2
81	96	5	52	80	4	23	63	2
80	95	5	51	80	4	22	61	2
79	94	5	50	80	4	21	60	2
78	94	5	49	79	4	20	58	2
77	93	5	48	79	4	19	55	1
76	92	5	47	79	4	18	54	1
75	91	5	46	78	4	17	53	1
74	91	5	45	78	4	16	51	1
73	90	5	44	77	3	15	49	1
72	89	5	43	77	3	14	47	1
71	89	5	42	77	3	13	44	1
70	88	5	41	76	3	12	42	1
69	88	5	40	76	3	11	39	1
68	87	5	39	76	3	10	37	1
67	87	5	38	75	3	9	34	1
66	86	5	37	75	3	8	31	1
65	86	5	36	74	3	7	27	1
64	85	5	35	73	3	6	24	1
63	84	4	34	73	3	5	20	1
62	84	4	33	72	3	4	17	1
61	84	4	32	72	3	3	13	1
60	83	4	31	71	3	2	9	1
59	83	4	30	70	3	1	4	1
58	82	4	29	69	3	0	0	1

Examination
August 2016
Algebra II

HIGH SCHOOL MATH REFERENCE SHEET

Conversions

1 inch = 2.54 centimeters	1 cup = 8 fluid ounces
1 meter = 39.37 inches	1 pint = 2 cups
1 mile = 5280 feet	1 quart = 2 pints
1 mile = 1760 yards	1 gallon = 4 quarts
1 mile = 1.609 kilometers	1 gallon = 3.785 liters
	1 liter = 0.264 gallon
1 kilometer = 0.62 mile	1 liter = 1000 cubic centimeters
1 pound = 16 ounces	
1 pound = 0.454 kilogram	
1 kilogram = 2.2 pounds	
1 ton = 2000 pounds	

Formulas

Triangle	$A = \dfrac{1}{2}bh$
Parallelogram	$A = bh$
Circle	$A = \pi r^2$
Circle	$C = \pi d$ or $C = 2\pi r$

Formulas (continued)

General Prisms $\qquad V = Bh$

Cylinder $\qquad V = \pi r^2 h$

Sphere $\qquad V = \dfrac{4}{3}\pi r^3$

Cone $\qquad V = \dfrac{1}{3}\pi r^2 h$

Pyramid $\qquad V = \dfrac{1}{3}Bh$

Pythagorean Theorem $\qquad a^2 + b^2 = c^2$

Quadratic Formula $\qquad x = \dfrac{-b \pm \sqrt{b^2 - 4ac}}{2a}$

Arithmetic Sequence $\qquad a_n = a_1 + (n-1)d$

Geometric Sequence $\qquad a_n = a_1 r^{n-1}$

Geometric Series $\qquad S_n = \dfrac{a_1 - a_1 r^n}{1 - r}$ where $r \neq 1$

Radians $\qquad$ 1 radian $= \dfrac{180}{\pi}$ degrees

Degrees $\qquad$ 1 degree $= \dfrac{\pi}{180}$ radians

Exponential Growth/Decay $\qquad A = A_0 e^{k(t - t_0)} + B_0$

PART I

Answer all 24 questions in this part. Each correct answer will receive 2 credits. No partial credit will be allowed. For each statement or question, write in the space provided the numeral preceding the word or expression that best completes the statement or answers the question. [48 credits]

1 Which equation has $1 - i$ as a solution?

(1) $x^2 + 2x - 2 = 0$ (3) $x^2 - 2x - 2 = 0$

(2) $x^2 + 2x + 2 = 0$ (4) $x^2 - 2x + 2 = 0$ 1 _____

2 Which statement(s) about statistical studies is true?

 I. A survey of all English classes in a high school would be a good sample to determine the number of hours students throughout the school spend studying.

 II. A survey of all ninth graders in a high school would be a good sample to determine the number of student parking spaces needed at that high school.

 III. A survey of all students in one lunch period in a high school would be a good sample to determine the number of hours adults spend on social media websites.

 IV. A survey of all Calculus students in a high school would be a good sample to determine the number of students throughout the school who don't like math.

(1) I, only (3) I and III

(2) II, only (4) III and IV 2 _____

3 To the *nearest tenth*, the value of x that satisfies $2^x = -2x + 11$ is

(1) 2.5   (3) 5.8
(2) 2.6   (4) 5.9   3 _____

4 The lifespan of a 60-watt lightbulb produced by a company is normally distributed with a mean of 1450 hours and a standard deviation of 8.5 hours. If a 60-watt lightbulb produced by this company is selected at random, what is the probability that its lifespan will be between 1440 and 1465 hours?

(1) 0.3803   (3) 0.8415
(2) 0.4612   (4) 0.9612   4 _____

5 Which factorization is *incorrect*?

(1) $4k^2 - 49 = (2k + 7)(2k - 7)$
(2) $a^3 - 8b^3 = (a - 2b)(a^2 + 2ab + 4b^2)$
(3) $m^3 + 3m^2 - 4m + 12 = (m - 2)^2(m + 3)$
(4) $t^3 + 5t^2 + 6t + t^2 + 5t + 6 = (t + 1)(t + 2)(t + 3)$ 5 _____

6 Sally's high school is planning their spring musical. The revenue, R, generated can be determined by the function $R(t) = -33t^2 + 360t$, where t represents the price of a ticket. The production cost, C, of the musical is represented by the function $C(t) = 700 + 5t$. What is the highest ticket price, to *the nearest dollar*, they can charge in order to *not* lose money on the event?

(1) $t = 3$ (3) $t = 8$
(2) $t = 5$ (4) $t = 11$ 6 _____

7 The set of data in the table below shows the results of a survey on the number of messages that people of different ages text on their cell phones each month.

Age Group	Text Messages per Month		
	0–10	11–50	Over 50
15–18	4	37	68
19–22	6	25	87
23–60	25	47	157

If a person from this survey is selected at random, what is the probability that the person texts over 50 messages per month given that the person is between the ages of 23 and 60?

(1) $\dfrac{157}{229}$ (3) $\dfrac{157}{384}$

(2) $\dfrac{157}{312}$ (4) $\dfrac{157}{456}$ 7 _____

8 A recursive formula for the sequence $18, 9, 4.5, \ldots$ is

(1) $g_1 = 18$ (3) $g_1 = 18$

     $g_n = \dfrac{1}{2} g_{n-1}$ $g_n = 2g_{n-1}$

(2) $g_n = 18\left(\dfrac{1}{2}\right)^{n-1}$ (4) $g_n = 18(2)^{n-1}$ 8 _____

9 Kristin wants to increase her running endurance. According to experts, a gradual mileage increase of 10% per week can reduce the risk of injury. If Kristin runs 8 miles in week one, which expression can help her find the total number of miles she will have run over the course of her 6-week training program?

(1) $\displaystyle\sum_{n=1}^{6} 8(1.10)^{n-1}$

(3) $\dfrac{8-8(1.10)^6}{0.90}$

(2) $\displaystyle\sum_{n=1}^{6} 8(1.10)^{n}$

(4) $\dfrac{8-8(0.10)^n}{1.10}$ 9 _____

10 A sine function increasing through the origin can be used to model light waves. Violet light has a wavelength of 400 nanometers. Over which interval is the height of the wave *decreasing*, only?

(1) $(0, 200)$

(3) $(200, 400)$

(2) $(100, 300)$

(4) $(300, 400)$ 10 _____

11 The expression $\dfrac{x^3 + 2x^2 + x + 6}{x+2}$ is equivalent to

(1) $x^2 + 3$

(3) $2x^2 + x + 6$

(2) $x^2 + 1 + \dfrac{4}{x+2}$

(4) $2x^2 + 1 + \dfrac{4}{x+2}$ 11 _____

12 A candidate for political office commissioned a poll. His staff received responses from 900 likely voters and 55% of them said they would vote for the candidate. The staff then conducted a simulation of 1000 more polls of 900 voters, assuming that 55% of voters would vote for their candidate. The output of the simulation is shown in the diagram below.

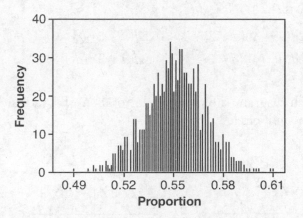

Given this output, and assuming a 95% confidence level, the margin of error for the poll is closest to

(1) 0.01   (3) 0.06
(2) 0.03   (4) 0.12   12 _____

13 An equation to represent the value of a car after t months of ownership is $v = 32{,}000(0.81)^{\frac{t}{12}}$. Which statement is *not* correct?

(1) The car lost approximately 19% of its value each month.
(2) The car maintained approximately 98% of its value each month.
(3) The value of the car when it was purchased was $32,000.
(4) The value of the car 1 year after it was purchased was $25,920.   13 _____

14 Which equation represents an odd function?

    (1) $y = \sin x$ (3) $y = (x + 1)^3$

    (2) $y = \cos x$ (4) $y = e^{5x}$ 14 _____

15 The completely factored form of
   $2d^4 + 6d^3 - 18d^2 - 54d$ is

    (1) $2d(d^2 - 9)(d + 3)$ (3) $2d(d + 3)^2(d - 3)$

    (2) $2d(d^2 + 9)(d + 3)$ (4) $2d(d - 3)^2(d + 3)$ 15 _____

16 Which diagram shows an angle rotation of 1 radian on the unit circle?

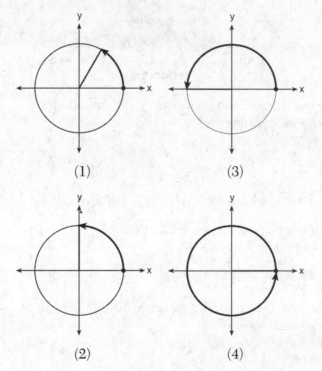

       (1)   (3)

       (2)   (4)   16 _____

17 The focal length, F, of a camera's lens is related to the distance of the object from the lens, J, and the distance to the image area in the camera, W, by the formula below.

$$\frac{1}{J} + \frac{1}{W} = \frac{1}{F}$$

When this equation is solved for J in terms of F and W, J equals

(1) $F - W$ (3) $\dfrac{FW}{W - F}$

(2) $\dfrac{FW}{F - W}$ (4) $\dfrac{1}{F} - \dfrac{1}{W}$ 17 _____

18 The sequence $a_1 = 6$, $a_n = 3a_{n-1}$ can also be written as

(1) $a_n = 6 \cdot 3^n$ (3) $a_n = 2 \cdot 3^n$

(2) $a_n = 6 \cdot 3^{n+1}$ (4) $a_n = 2 \cdot 3^{n+1}$ 18 _____

19 Which equation represents the set of points equidistant from line l and point R shown on the graph below?

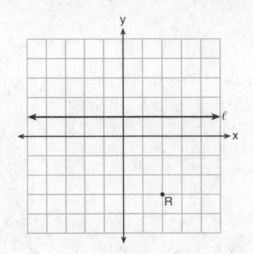

(1) $y = -\dfrac{1}{8}(x + 2)^2 + 1$

(3) $y = -\dfrac{1}{8}(x - 2)^2 + 1$

(2) $y = -\dfrac{1}{8}(x + 2)^2 - 1$

(4) $y = -\dfrac{1}{8}(x - 2)^2 - 1$

19 _____

20 Mr. Farison gave his class the three mathematical rules shown below to either prove or disprove. Which rules can be proved for all real numbers?

$$I \quad (m + p)^2 = m^2 + 2mp + p^2$$
$$II \quad (x + y)^3 = x^3 + 3xy + y^3$$
$$III \quad (a^2 + b^2)^2 = (a^2 - b^2)^2 + (2ab)^2$$

(1) I, only

(3) II and III

(2) I and II

(4) I and III

20 _____

21 The graph of $p(x)$ is shown below.

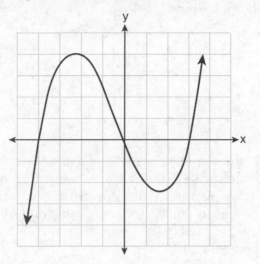

What is the remainder when $p(x)$ is divided by $x + 4$?

(1) $x - 4$ (3) 0

(2) -4 (4) 4   21 _____

22 A payday loan company makes loans between $100 and $1000 available to customers. Every 14 days, customers are charged 30% interest with compounding. In 2013, Remi took out a $300 payday loan. Which expression can be used to calculate the amount she would owe, in dollars, after one year if she did not make payments?

(1) $300(.30)^{\frac{14}{365}}$

(2) $300(1.30)^{\frac{14}{365}}$

(3) $300(.30)^{\frac{365}{14}}$

(4) $300(1.30)^{\frac{365}{14}}$ 22 _____

23 Which value is *not* contained in the solution of the system shown below?

$$a + 5b - c = -20$$
$$4a - 5b + 4c = 19$$
$$-a - 5b - 5c = 2$$

(1) –2   (3) 3

(2) 2   (4) –3   23 _____

24 In 2010, the population of New York State was approximately 19,378,000 with an annual growth rate of 1.5%. Assuming the growth rate is maintained for a large number of years, which equation can be used to predict the population of New York State t years after 2010?

(1) $P_t = 19,378,000(1.5)t$

(2) $P_0 = 19,378,000$
     $P_t = 19,378,000 + 1.015P_{t-1}$

(3) $P_t = 19,378,000\ (1.015)^{t-1}$

(4) $P_0 = 19,378,000$
     $P_t = 1.015P_{t-1}$ 24 _____

PART II

Answer all 8 questions in this part. Each correct answer will receive 2 credits. Clearly indicate the necessary steps, including appropriate formula substitutions, diagrams, graphs, charts, etc. For all questions in this part, a correct numerical answer with no work shown will receive only 1 credit.   [16 credits]

25 The volume of air in a person's lungs, as the person breathes in and out, can be modeled by a sine graph. A scientist is studying the differences in this volume for people at rest compared to people told to take a deep breath. When examining the graphs, should the scientist focus on the amplitude, period, or midline? Explain your choice.

26 Explain how $\left(3^{\frac{1}{5}}\right)^{2}$ can be written as the equivalent radical expression $\sqrt[5]{9}$.

27 Simplify $xi(i - 7i)^2$, where i is the imaginary unit.

28 Using the identity $\sin^2 \theta + \cos^2 \theta = 1$, find the value of $\tan \theta$, to the *nearest hundredth*, if $\cos \theta$ is -0.7 and θ is in Quadrant II.

29 Elizabeth waited for 6 minutes at the drive thru at her favorite fast-food restaurant the last time she visited. She was upset about having to wait that long and notified the manager. The manager assured her that her experience was very unusual and that it would not happen again.

A study of customers commissioned by this restaurant found an approximately normal distribution of results. The mean wait time was 226 seconds and the standard deviation was 38 seconds. Given these data, and using a 95% level of confidence, was Elizabeth's wait time unusual? Justify your answer.

30 The x-value of which function's x-intercept is larger, f or h? Justify your answer.

$f(x) = \log(x - 4)$

x	h(x)
−1	6
0	4
1	2
2	0
3	−2

31 The distance needed to stop a car after applying the brakes varies directly with the square of the car's speed. The table below shows stopping distances for various speeds.

Speed (mph)	10	20	30	40	50	60	70
Distance (ft)	6.25	25	56.25	100	156.25	225	306.25

Determine the average rate of change in braking distance, in ft/mph, between one car traveling at 50 mph and one traveling at 70 mph.

Explain what this rate of change means as it relates to braking distance.

32 Given events A and B, such that $P(A) = 0.6$, $P(B) = 0.5$, and $P(A \cup B) = 0.8$, determine whether A and B are independent or dependent.

PART III

Answer all 4 questions in this part. Each correct answer will receive 4 credits. Clearly indicate the necessary steps, including appropriate formula substitutions, diagrams, graphs, charts, etc. For all questions in this part, a correct numerical answer with no work shown will receive only 1 credit. [16 credits]

33 Find algebraically the zeros for $p(x) = x^3 + x^2 - 4x - 4$.

On the set of axes below, graph $\hat{y} = p(x)$.

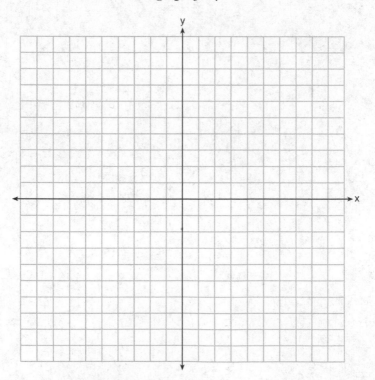

34 One of the medical uses of Iodine–131 (I–131), a radioactive isotope of iodine, is to enhance x-ray images. The half-life of I–131 is approximately 8.02 days. A patient is injected with 20 milligrams of I–131. Determine, to the *nearest day*, the amount of time needed before the amount of I–131 in the patient's body is approximately 7 milligrams.

35 Solve the equation $\sqrt{2x-7} + x = 5$ algebraically, and justify the solution set.

36 Ayva designed an experiment to determine the effect of a new energy drink on a group of 20 volunteer students. Ten students were randomly selected to form group 1 while the remaining 10 made up group 2. Each student in group 1 drank one energy drink, and each student in group 2 drank one cola drink. Ten minutes later, their times were recorded for reading the same paragraph of a novel. The results of the experiment are shown below.

Group 1 (seconds)	Group 2 (seconds)
17.4	23.3
18.1	18.8
18.2	22.1
19.6	12.7
18.6	16.9
16.2	24.4
16.1	21.2
15.3	21.2
17.8	16.3
19.7	14.5
Mean = 17.7	Mean = 19.1

a) Ayva thinks drinking energy drinks makes students read faster. Using information from the experimental design or the results, explain why Ayva's hypothesis may be *incorrect*.

Question 36 continued on next page.

Question 36 continued.

Using the given results, Ayva randomly mixes the 20 reading times, splits them into two groups of 10, and simulates the difference of the means 232 times.

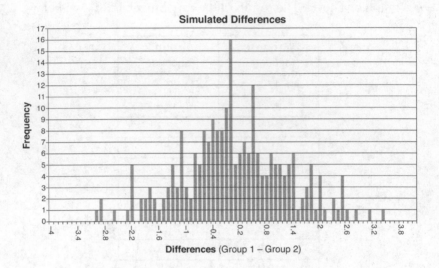

b) Ayva has decided that the difference in mean reading times is *not* an unusual occurence. Support her decision using the results of the simulation. Explain your reasoning.

PART IV

Answer the question in this part. A correct answer will receive 6 credits. Clearly indicate the necessary steps, including appropriate formula substitutions, diagrams, graphs, charts, etc. A correct numerical answer with no work shown will receive only 1 credit. [6 credits]

37 Seth's parents gave him $5000 to invest for his 16th birthday. He is considering two investment options. Option A will pay him 4.5% interest compounded annually. Option B will pay him 4.6% compounded quarterly.

Write a function of option A and option B that calculates the value of each account after n years.

Seth plans to use the money after he graduates from college in 6 years. Determine how much more money option B will earn than option A to the *nearest cent*.

Algebraically determine, to the *nearest tenth of a year*, how long it would take for option B to double Seth's initial investment.

Answers
August 2016
Algebra II

Answer Key

PART I

1. (4)	**5.** (3)	**9.** (1)	**13.** (1)	**17.** (3)	**21.** (3)
2. (1)	**6.** (3)	**10.** (2)	**14.** (1)	**18.** (3)	**22.** (4)
3. (2)	**7.** (1)	**11.** (2)	**15.** (3)	**19.** (4)	**23.** (2)
4. (3)	**8.** (1)	**12.** (2)	**16.** (1)	**20.** (4)	**24.** (4)

PART II

25. Amplitude

26. $\sqrt[5]{9} = 9^{\frac{1}{5}} = \left(3^2\right)^{\frac{1}{5}} = 3^{\frac{2}{5}} = \left(3^{\frac{1}{5}}\right)^2$

27. $-36xi$

28. -1.02

29. Yes, Elizabeth's wait time was unusual.

30. f

31. 7.5

32. Independent

PART III

33. $2, -2, -1$

34. 12 days

35. 4

36. (a) The hypothesis might be incorrect.

(b) The difference is not an unusual occurrence.

PART IV

37. $A(x) = 5000\left(1 + \dfrac{0.045}{1}\right)^{1x}$;

$B(x) = 5000\left(1 + \dfrac{0.046}{4}\right)^{4x}$;

$67.57; 15.2 years

In **Parts II–IV**, you are required to show how you arrived at your answers. For sample methods of solutions, see the *Answers Explained* section.

Answers Explained

PART I

1. Use the quadratic formula, $x = \dfrac{-b \pm \sqrt{b^2 - 4ac}}{2a}$, with each choice.

Choice (1):

$$x = \frac{-2 \pm \sqrt{2^2 - 4 \cdot 1 \cdot (-2)}}{2 \cdot 1}$$

$$= \frac{-2 \pm \sqrt{4 + 8}}{2}$$

$$= \frac{-2 \pm \sqrt{12}}{2}$$

$$= \frac{-2 \pm 2\sqrt{3}}{2}$$

$$= -1 \pm \sqrt{3}$$

Choice (2):

$$x = \frac{-2 \pm \sqrt{2^2 - 4 \cdot 1 \cdot 2}}{2 \cdot 1}$$

$$= \frac{-2 \pm \sqrt{4 - 8}}{2}$$

$$= \frac{-2 \pm \sqrt{-4}}{2}$$

$$= \frac{-2 \pm 2i}{2}$$

$$= -1 \pm i$$

Choice (3):

$$x = \frac{-(-2) \pm \sqrt{(-2)^2 - 4 \cdot 1 \cdot (-2)}}{2 \cdot 1}$$

$$= \frac{2 \pm \sqrt{4 + 8}}{2}$$

$$= \frac{2 \pm \sqrt{12}}{2}$$

$$= \frac{2 \pm 2\sqrt{3}}{2}$$

$$= 1 \pm \sqrt{3}$$

Choice (4):

$$x = \frac{-(-2) \pm \sqrt{(-2)^2 - 4 \cdot 1 \cdot 2}}{2 \cdot 1}$$

$$= \frac{2 \pm \sqrt{4 - 8}}{2}$$

$$= \frac{2 \pm \sqrt{-4}}{2}$$

$$= \frac{2 \pm 2i}{2}$$

$$= 1 \pm i$$

The correct choice is **(4)**.

2. Check each of the four statements.

Check I: The most accurate sample you can use for a survey is the entire population. Since this will be a survey of all English classes, it is a good sample.

Check II: Since ninth graders do not drive cars, they are not a good sample to survey about parking spaces.

Check III: High school students won't know how many hours adults spend on social media. A better option would be to use a random sample that includes adults.

Check IV: Using a random sample of all students rather than just choosing students who are taking Calculus would be better. The students in Calculus probably like math more than students who are not in Calculus, so the results of a survey using only Calculus will be skewed.

The correct choice is **(1)**.

3. There is not a simple way to solve this equation using just Algebra. One way to solve this would be to use the graphing calculator to graph $y = 2^x$ and $y = -2x + 11$ on the same set of axes. The x-coordinate of the intersection point is the solution to the equation. For this example, that x-coordinate is approximately 2.557, which rounds to 2.6.

For the TI-84:

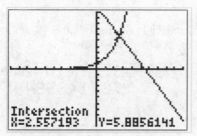

For the TI-Nspire:

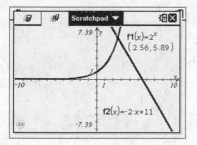

Another way would be to test the four choices to see which makes 2^x closest to $-2x + 11$. For $x = 2.6$, $2^{2.6} \approx 6.1$ while $-2 \cdot 2.6 + 11 = 5.8$.

The correct choice is **(2)**.

4. This question requires you use the normal cdf function of the graphing calculator. Four numbers need to be entered into the function. These are the mean (1450), the standard deviation (8.5), the lower bound (1440), and the upper bound (1465). The calculator will use these numbers to calculate the probability that a randomly chosen bulb will have a lifespan between 1440 and 1465 hours.

For the TI-84:

Press [2ND] [VARS] [2] for normalcdf. Fill in the fields with 1440 for lower, 1465 for upper, 1450 for μ, and 8.5 for σ. Then select "Paste."

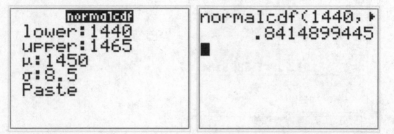

For the TI-Nspire:

From the home screen, press [A]. Press [menu] [6] [5] [2] for Normal Cdf. Set the Lower Bound to 1440, the Upper Bound to 1465, the mean (μ) to 1450, and the standard deviation (σ) to 8.5.

The probability is approximately 0.8415.

The correct choice is **(3)**.

5. Factor each of the answer choices to find out which one is incorrect.

Choice (1): This choice uses the difference of two squares factoring pattern.

$$4k^2 - 49 = (2k)^2 - 7^2 = (2k + 7)(2k - 7)$$

Choice (2): This choice can be checked by multiplying the expression on the right side of the equal sign to see if it simplifies to the expression on the left side of the equal sign.

$$(a - 2b)(a^2 + 2ab + 4b^2)$$
$$= a \cdot a^2 + a \cdot 2ab + a \cdot 4b^2 + (-2b)a^2 + (-2b)2ab + (-2b)4b^2$$
$$= a^3 + 2a^2b + 4ab^2 - 2a^2b - 4ab^2 - 8b^3 = a^3 - 8b^3$$

Choice (3): This choice can be checked by multiplying the expression on the right side of the equal to sign to see if it simplifies to the expression on the left side of the equal sign.

$$(m - 2)^2(m + 3) = (m - 2)(m - 2)(m + 3) = (m^2 - 4m + 4)(m + 3)$$
$$= m^2 \cdot m + m^2 \cdot 3 + (-4m)m + (-4m)3 + 4 \cdot m + 4 \cdot 3$$
$$= m^3 + 3m^2 - 4m^2 - 12m + 4m + 12 = m^3 - m^2 - 8m + 12$$

This does not equal the expression on the left side of the equal sign.

Choice (4): The expression on the left side of the equal sign can be factored by grouping.

$$t^3 + 5t^2 + 6t + t^2 + 5t + 6 = t(t^2 + 5t + 6) + 1(t^2 + 5t + 6)$$
$$= (t + 1)(t^2 + 5t + 6) = (t + 1)(t + 2)(t + 3)$$

The correct choice is **(3)**.

6. The high school will not lose money when the revenue minus the cost is greater than 0. Use an algebraic inequality to model this situation:

$$R(t)^3 - C(t) = (-33t^2 + 360t) - (700 + 5t)$$
$$= -33t + 360t - 700 - 5t$$
$$= -33t^2 + 355t - 700 \geq 0$$

Algebraically solving a quadratic inequality like this can be very difficult. Fortunately, this is a multiple-choice question. Since you want to know which answer choice is the highest price that can be charged without losing money on the event, start by seeing if choice (4) makes the inequality true. It is the highest number.

Choice (4): $-33 \cdot 11^2 + 355 \cdot 11 - 700 = -788$. If the ticket price is $11, the high school will lose $788.

Choice (3): $-33 \cdot 8^2 + 355 \cdot 8 - 700 = +28$. If the ticket price is \$8, the high school will lose \$28.

Choices (1) and (2) also result in a positive value. Since those choices are lower than choice (3) and the question asks for the highest ticket price that will not lose money, $t = 8$ is the answer.

The correct choice is **(3)**.

7. It is "given" that the person is between the ages of 23 and 60. So add together the numbers in the last row, $25 + 47 + 157 = 229$, to get the denominator of the fraction. Exactly 157 people who are between the ages of 23 and 60 send over 50 text messages per month. So the probability that a randomly selected person texts over 50 messages per month given that the person is between the ages of 23 and 60 is $\dfrac{157}{229}$.

The correct choice is **(1)**.

8. A recursive formula first defines one or more terms of the sequence and then describes a formula for obtaining other terms by substituting values from terms earlier in the sequence into the recursive part of the formula. Since choices (2) and (4) are direct formulas, only choices (1) and (3) are recursive formulas. Check choices (1) and (3) to see if they result in the given sequence.

Choice (1): $g_1 = 18, g_2 = \dfrac{1}{2} g_{2-1} = \dfrac{1}{2} g_1 = \dfrac{1}{2} \cdot 18 = 9$

This agrees with the first two terms of the sequence.

Choice (3): $g_1 = 18, g_2 = 2g_{2-1} = 2g_1 = 2 \cdot 18 = 36$

This agrees with the first term of the sequence but not with the second term.

Choice (2) is a direct formula for the given sequence. However, the question specifically asks for the recursive formula.

The correct choice is **(1)**.

9. In week 1, Kristin runs 8 miles. In week 2, she runs 10% more. Since 10% of 8 is 0.8, Kristin runs $8 + 0.8 = 8.8$ miles in week 2. Another way to calculate the number of miles Kristin runs in the second week is to multiply the number of miles she ran the first week by 1.1 to get $8 \cdot 1.1 = 8.8$ miles. To find the number of miles she runs in week 3, multiply the number of miles she ran in the second week by 1.1, which gives $8.8 \cdot 1.1 = 9.68$. Continuing this pattern and multiplying each number by 1.1 to get the next number generates the following series:

$$8 + 8.8 + 9.68 + 10.648 + 11.7128 + 12.88408$$

The total number of miles Kristin runs in 6 weeks is the sum of these numbers.

Choice (1) can be expanded by beginning with $n = 1$ and ending with $n = 6$ to become the following:

For $n = 1$: $8(1.10)^{1-1} - 8(1.10)^0 - 8(1) = 8$

For $n = 2$: $8(1.10)^{2-1} - 8(1.10)^1 - 8(1.10) = 8.8$

For $n = 3$: $8(1.10)^{3-1} - 8(1.10)^2 - 8(1.21) = 9.68$

Inserting $n = 4, 5,$ and 6 will match the numbers in the series calculated above.

Choice (2) would create a similar series but starting with 8.8 and ending with 14.172488 instead of 12.88408. So this choice does not match the series calculated above.

Choices (3) and (4) are not correct. However, there is an expression that resembles those choices that would be correct if it were an option. This is because the series can be written as the geometric series $8 + 8(1.1) + 8(1.1)^2 + 8(1.1)^3 + 8(1.1)^4 + 8(1.1)^5$. The sum of a geometric series is given in the reference sheet in the back of the Regents booklet:

$$S_n = \frac{a_1 - a_1 r^n}{1-r}$$

For this example, $a_1 = 8$, $r = 1.1$, and $n = 6$, so

$$S_n = \frac{8 - 8(1.10)^6}{1-1.1} = \frac{8 - 8(1.10)^6}{-0.10}$$

This looks similar to choice (3) but not exactly.

The correct choice is **(1)**.

10. A sine curve with a period of 400 is graphed below. As a point moves on the curve from left to right, the value is increasing if the graph is going up and is decreasing if the graph is going down. This curve is increasing from 0 to 100, decreasing from 100 to 300, and then increasing again from 300 to 400.

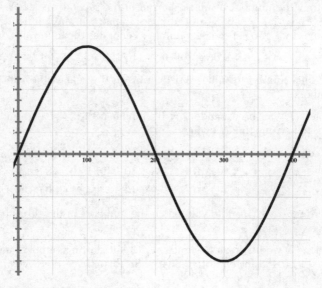

The correct choice is (**2**).

11. Use polynomial long division:

$$x + 2 \overline{\smash{)}\, x^3 + 2x^2 + x + 6} \quad \begin{array}{l} x^2 + 0x + 1 + \dfrac{4}{x+2} \end{array}$$

$$
\begin{array}{r}
x^2 + 0x + 1 + \dfrac{4}{x+2} \\
x + 2 \,\overline{\smash{)}\, x^3 + 2x^2 + x + 6} \\
-\left(x^3 + 2x^2\right) \\
\hline
0x^2 + x \\
-\left(0x^2 + 0x\right) \\
\hline
x + 6 \\
-\left(x + 2\right) \\
\hline
4
\end{array}
$$

The correct choice is (**2**).

12. For a 95% confidence interval, nearly all of the values will be between the mean plus 2 times the margin of error and the mean minus 2 times the margin of error. The mean in this case is 0.55. Since the maximum value is about 0.06 above the mean and the minimum value is about 0.06 below the mean, the margin of error is $\dfrac{0.06}{2} = 0.03$.

The correct choice is **(2)**.

13. Test the four answer choices.

Choice (1): When an equation is in the form $y = a(1-r)^x$, the variable r is the percent decrease. This equation can be rewritten:

$$v = 32{,}000\left((0.81)^{\frac{1}{12}}\right)^x = 32{,}000(0.98)^x = 32{,}000(1-0.02)^x$$

So the car loses approximately 2% of its value each month, not 19%.

Choice (2): Rewrite the equation:

$$v = 32{,}000\left((0.81)^{\frac{1}{12}}\right)^x = 32{,}000(0.98)^x$$

So the car does retain 98% of its value each month.

Choice (3): The value of the car when it was purchased can be calculated by substituting $t = 0$:

$$v = 32{,}000(0.81)^{\frac{0}{12}} = 32{,}000(0.81)^0 = 32{,}000(1) = 32{,}000$$

Choice (4): To find the value of the car 1 year after purchasing it, substitute $t = 12$ since t is measured in months:

$$v = 32{,}000\,(0.81)^{\frac{12}{12}} = 32{,}000(0.81)^1 = 32{,}000(0.81) = 25{,}920$$

The correct choice is **(1)**.

14. The graph of an odd function is symmetric with respect to the origin. Here are the graphs for the four functions in the answer choices:

Choice (1):

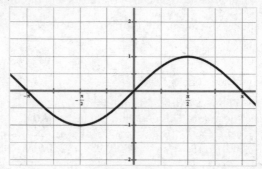

Choice (2):

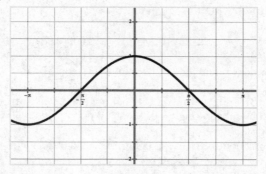

Choice (3):

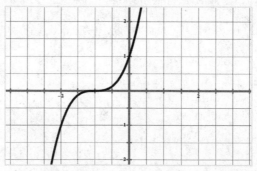

Choice (4):

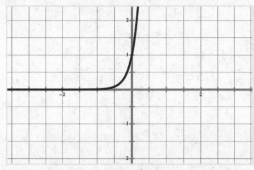

Only choice (1) is symmetric with respect to the origin. Choice (2) is an even function because it is symmetric with respect to the y-axis. Choice (3) is symmetric with respect to the point $(-1, 0)$, which is not the origin. Choice (4) does not have any obvious symmetry.

The correct choice is **(1)**.

15. This expression can be factored by grouping by using the and difference of perfect squares:

$$2d^4 + 6d^3 - 18d^2 - 54d$$
$$2d^3(d+3) - 18d(d+3)$$
$$(2d^3 - 18d)(d+3)$$
$$2d(d^2 - 9)(d+3)$$
$$2d(d-3)(d+3)(d+3)$$
$$2d(d-3)(d+3)^2$$

Instead of factoring, you can answer this question by multiplying out each of the answer choices to see which one is equivalent to the given equation.

The correct choice is **(3)**.

16. The formula for converting radians to degrees is degrees = radians $\cdot \dfrac{180}{\pi}$.

Using this formula, the number of degrees in 1 radian is equal to approximately 57.3 degrees. Only in choice (1) is the angle of rotation less than 90 degrees.

The correct choice is **(1)**.

17. To solve for J in terms of W and F, start by isolating the term that contains J:

$$\frac{1}{J} + \frac{1}{W} = \frac{1}{F}$$

$$-\frac{1}{W} = -\frac{1}{W}$$

$$\frac{1}{J} = \frac{1}{F} - \frac{1}{W}$$

Now simplify the right side of the equation by using the common denominator FW:

$$\frac{1}{J} = \frac{1}{F} - \frac{1}{W}$$

$$\frac{1}{J} = \frac{W}{FW} - \frac{F}{FW}$$

$$\frac{1}{J} = \frac{W - F}{FW}$$

Finally, take the reciprocal of both sides so the left side becomes J:

$$\frac{1}{J} = \frac{W - F}{FW}$$

$$\frac{J}{1} = \frac{FW}{W - F}$$

The correct choice is **(3)**.

18. The terms of the sequence can first be determined with the recursive formula. Because $a_1 = 6$, the first term is 6. When $n = 2$, use the recursive part of the definition:

$$a_2 = 3a_{2-1} = 3a_1 = 3 \cdot 6 = 18$$

For $n = 3$, again use the recursive part of the definition:

$$a_3 = 3a_{3-1} = 3a_2 = 3 \cdot 18 = 54$$

Each term becomes 3 times the previous term to create the sequence 6, 18, 54, 162, 486, . . .

The answer choices are all direct formulas for calculating the terms of a sequence. Substitute $n = 1$, 2, and 3 into each formula to see which one leads to a sequence starting as 6, 18, and 54.

Choice (1): For $n = 1$:

$$a_1 = 6 \cdot 3^1 = 6 \cdot 3 = 18 \neq 6$$

Choice (2): For $n = 1$:

$$a_1 = 6 \cdot 3^{1+1} = 6 \cdot 3^2 = 6 \cdot 9 = 54 \neq 6$$

Choice (3): For $n = 1$, $n = 2$, and $n = 3$:

$$a_1 = 2 \cdot 3^1 = 2 \cdot 3 = 6$$
$$a_2 = 2 \cdot 3^2 = 2 \cdot 9 = 18$$
$$a_3 = 2 \cdot 3^3 = 2 \cdot 27 = 54$$

Choice (4): For $n = 1$:

$$a_1 = 2 \cdot 3^{1+1} = 2 \cdot 3^2 = 2 \cdot 9 = 18 \neq 6$$

The correct choice is **(3)**.

19. The set of points equidistant from a line and a point is a parabola that has the line as its directrix and the point as its focus. The vertex of the parabola is halfway between the focus and the directrix. Since the focus in this question is 4 units below the directrix, the vertex of the parabola will be 2 units above point R. So the vertex has the coordinates $(2, -1)$.

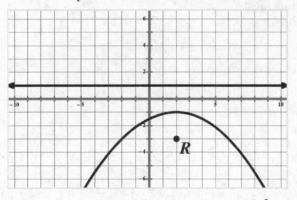

All four answer choices are in vertex form $y = a(x - h)^2 + k$, where the coordinates of the vertex are (h, k). Since the a-value for each of the answer choices is the same, the only difference is the vertex.

Since the vertex is $(2, -1)$, $h = 2$ and $k = -1$. So the equation must be in the form

$$y = a(x - 2)^2 + (-1) = a(x - 2)^2 - 1$$

The correct choice is **(4)**.

20. The left and right sides of each equation should be simplified to see if they are equivalent.

Check I:

$$(m + p)^2 = (m + p)(m + p) = m^2 + mp + mp + p^2 = m^2 + 2mp + p^2$$

This rule can be proved for all real numbers.

Check II:

$$(x + y)^3 = (x + y)(x + y)(x + y) = (x^2 + 2xy + y^2)(x + y)$$
$$= x^2 \cdot x + x^2 \cdot y + 2xy \cdot x + 2xy \cdot y + y^2 \cdot x + y^2 \cdot y$$
$$= x^3 + x^2y + 2x^2y + 2xy^2 + xy^2 + y^3$$
$$= x^3 + 3x^2y + 3xy^2 + y^3 \neq x^3 + 3xy + y^3$$

This rule cannot be proved for all real numbers.

Check III:

First simplify the left side:

$$(a^2 + b^2)^2 = (a^2 + b^2)(a^2 + b^2) = a^2 \cdot a^2 + a^2 \cdot b^2 + a^2 \cdot b^2 + b^2 \cdot b^2$$
$$= a^4 + 2a^2b^2 + b^4$$

Then simplify the right side:

$$(a^2 - b^2)^2 + (2ab)^2 = (a^2 - b^2)(a^2 - b^2) + (2ab)^2$$
$$= a^2 \cdot a^2 - a^2 \cdot b^2 - a^2 \cdot b^2 + b^2 \cdot b^2 + 4a^2b^2$$
$$= a^4 - 2a^2b^2 + b^4 + 4a^2b^2 = a^4 + 2a^2b^2 + b^4$$

This rule can be proved for all real numbers.

The correct choice is **(4)**.

21. Because the graph of $p(x)$ has x-intercepts at –4, 0, and 3, the equation has the form $p(x) = a(x + 4)(x)(x - 3)$. Since $(x + 4)$ is a factor of $p(x)$, when you divide $p(x)$ by $(x + 4)$, the remainder will be 0.

This question can also be answered with the remainder theorem. The remainder theorem states that the remainder when $p(x)$ is divided by $(x - a)$ will be equal to $p(a)$. For this example, the remainder when $p(x)$ is divided by $(x + 4) = (x - (-4))$ will be equal to $p(-4)$. According to the graph, $p(-4) = 0$. So the remainder will also be 0.

The correct choice is **(3)**.

22. The easiest way to solve this question is to calculate the numerical value of each of the choices to see which is the most reasonable answer. With interest of 30% compounding every 14 days, the amount owed after a year should be a lot more than the original $300.

Choice (1):

$$300(.30)^{\frac{14}{365}} = 286.46$$

Choice (2):

$$300(1.30)^{\frac{14}{365}} = 303.03$$

Choice (3):

$$300(.30)^{\frac{365}{14}} \approx 0$$

Choice (4):

$$300(1.30)^{\frac{365}{14}} = 280,405.95$$

Even though choice (4) seems unbelievably high, all the other choices are way too low. Choice (3), which is nearly 0, is especially low. Choice (1) is actually lower than the original loan of $300.

The correct choice is **(4)**.

23. This system of three equations in three unknowns can be solved using the elimination method. Notice that the b-term in the first equation is +5 while the b-term in the second and third equations is –5.

Add the first and second equations to get a new equation with no b-term:

$$\begin{array}{r} a+5b-c=-20 \\ +\ \ 4a-5b+4c=19 \\ \hline 5a+3c=-1 \end{array}$$

Add the first and third equations to make another new equation with no b-term:

$$\begin{array}{r} a+5b-c=-20 \\ +\ \ -a-5b-5c=2 \\ \hline -6c=-18 \end{array}$$

In the second new equation, the a-term also got eliminated. Solve for c by dividing both sides of the equation by –6:

$$\frac{-6c}{-6} = \frac{-18}{-6}$$
$$c = 3$$

Substitute 3 for c into the first new equation, $5a + 3c = -1$, to solve for a:

$$5a + 3 \cdot 3 = -1$$
$$5a + 9 = -1$$
$$\underline{-9 = -9}$$
$$5a = -10$$
$$\frac{5a}{5} = \frac{-10}{5}$$
$$a = -2$$

Substitute 3 for c and –2 for a into any of the original equations to solve for b:

$$a + 5b - c = -20$$
$$-2 + 5b - 3 = -20$$
$$5b - 5 = -20$$
$$\underline{+5 = +5}$$
$$5b = -15$$
$$\frac{5b}{5} = \frac{-15}{5}$$
$$b = -3$$

The correct choice is (**2**).

24. You are given that the population in 2010 was 19,378,000. So $P_0 = 19,378,000$. After 1 year, the population increased by 1.5%. To calculate the increase in population, multiply 19,378,000 by 0.015. Then add that amount to the original population to find the population after 1 year, P_1:

$$19,378,000 \cdot 0.015 = 290,670$$
$$19,378,000 + 290,670 = 19,668,670$$
$$P_1 = 19,668,670$$

Test each answer choice to see which one has $P_0 = 19,378,000$ and $P_1 = 19,668,670$.

Choice (1):

$$P_0 = 19{,}378{,}000(1.5)^0 = 19{,}378{,}000 \cdot 1 = 19{,}378{,}000$$

$$P_1 = 19{,}378{,}000(1.5)^1 = 19{,}378{,}000 \cdot 1.5 = 29{,}067{,}000$$

Since the value of P_1 is wrong, this equation is not the correct answer.

Choice (2):

$$P_0 = 19{,}378{,}000$$

$$P_1 = 19{,}378{,}000 + 1.015P_{1-1} = 19{,}378{,}000 + 1.015P_0$$

$$= 19{,}378{,}000 + 1.015 \cdot 19{,}378{,}000 = 19{,}378{,}000 + 19{,}668{,}670$$

$$= 39{,}046{,}670$$

Since the value of P_1 is wrong, this equation is not the correct answer.

Choice (3):

$$P_0 = 19{,}378{,}000(1.015)^{0-1} = 19{,}378{,}000(1.015)^{-1}$$

$$= 19{,}378{,}000 \cdot 0.9852216749 \approx 19{,}091{,}625$$

Since the value of P_0 is wrong, this equation is not the correct answer.

Choice (4):

$$P_0 = 19{,}378{,}000$$

$$P_1 = 1.015P_{1-1} = 1.015P_0 = 1.015 \cdot 19{,}378{,}000$$

$$= 19{,}668{,}670$$

Since the values of P_0 and P_1 are correct, this equation is the correct answer.

A shorter way to solve this question is to know that when something grows by 1.5%, it gets multiplied by $1 + 0.015 = 1.015$. Choice (4) is a recursive sequence in which each term is equal to 1.015 multiplied by the previous term. This is the same as what happens in the model.

The correct choice is **(4)**.

PART II

25. When a person inhales, the volume of air in his or her lungs will increase until it reaches a maximum. When that person exhales, the volume of air in his or her lungs will decrease until it reaches a minimum.

Since breathing alternates between inhaling and exhaling, a graph of this scenario would look like this.

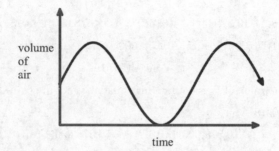

The amplitude of the sine curve is the difference between the maximum value and a middle value. Depending on the person, this amplitude can be large if the person's lungs can hold a lot of air or small if the person's lungs cannot hold a lot of air.

The scientist should focus on the amplitude.

26. $\left(3^{\frac{1}{5}}\right)^2$ can be simplified to $3^{\frac{1}{5}\cdot 2} = 3^{\frac{2}{5}}$.

$\sqrt[5]{9}$ is equivalent to $9^{\frac{1}{5}}$. Since $9 = 3^2$, $9^{\frac{1}{5}} = \left(3^2\right)^{\frac{1}{5}} = 3^{\frac{2}{5}} = \left(3^{\frac{1}{5}}\right)^2$.

27. $xi(i - 7i)^2 = xi(-6i)^2 = xi \cdot 36i^2 = xi \cdot 36(-1) = -36xi$.

The key step is replacing i^2 with -1 since that is how i is defined.

So $xi(i - 7i)^2$ simplifies to $-36xi$.

28. If $\cos \theta = -0.7$, start by substituting it into the given formula to solve for $\sin \theta$.

$$\sin^2 \theta + (-0.7)^2 = 1$$
$$\sin^2 \theta + 0.49 = 1$$
$$\underline{-0.49 = -0.49}$$
$$\sin^2 \theta = 0.51$$
$$\sqrt{\sin^2 \theta} = \pm\sqrt{0.51}$$
$$\sin \theta \approx \pm 0.7141428$$

Sine is positive only in quadrants I and II, while cosine is positive only in quadrants I and IV. Since it is given that θ is in quadrant II and since sine is positive in quadrant II, use the positive solution $\sin \theta \approx +0.7141428$. Now solve for $\tan \theta$ to the *nearest hundreth*:

$$\tan \theta = \frac{\sin \theta}{\cos \theta}$$

$$\tan \theta \approx \frac{0.7141428}{-0.7} \approx -1.02$$

29. This question is asking whether or not 6 minutes, or 360 seconds, is more than 2 standard deviations above the mean wait time.

You are given that the mean time is 226 seconds and the standard deviation is 38 seconds. A wait time more than 2 standard deviations, which is 76 seconds, above the mean or less than 2 standard deviations below the mean should be unusual.

Expected wait times are between 226 − 76 = 150 seconds and 226 + 76 = 302 seconds. Six minutes, which is 360 seconds, is greater than 302 seconds. So a wait time of 6 minutes is more than 2 standard deviations above the mean.

Elizabeth's wait time was unusual.

30. An x-intercept is the x-coordinate of the point on a graph that has a y-coordinate of 0. For the graph of $f(x)$, the x-intercept can be found by solving the equation:

$$0 = \log_{10}(x-4)$$
$$10^0 = x-4$$
$$1 = x-4$$
$$\underline{+4 = +4}$$
$$5 = x$$

So the x-intercept for $f(x)$ is 5.

According to the chart for $h(x)$, $h(2) = 0$. So the x-intercept for $h(x)$ is 2.

The x-intercept for function f is greater than the x-intercept for function h.

Function f has an x-intercept of 5. Function h has an x-intercept of 2.

31. The average rate of change can be calculated by dividing the change in braking distance divided by the change in speed. For this question, the calculation becomes:

$$\text{average rate of change} = \frac{306.25 - 156.25}{70 - 50} = \frac{150}{20} = 7.5$$

This rate of change means that for every 1 mile per hour above 50 miles per hour, the braking distance increases, on average, by about 7.5 feet.

32. Events are independent if $P(A \cap B) = P(A) \cdot P(B)$.

To check if events are independent in this question, organize the information on a Venn diagram. In the diagram below, $P(A)$ is the sum of the regions labeled II and III; $P(B)$ is the sum of the two regions labeled III and IV; $P(A \cup B)$ is the sum of the three regions labeled II, III, and IV; and $P(A \cap B)$ is the overlapping region labeled III.

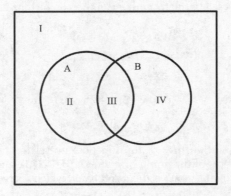

To find the probability for region III, $P(A \cup B)$, notice that

$$P(A) + P(B) = 1.1 = II + III + III + IV$$

Also it is given that $P(A \cup B) = 0.8 = II + III + IV$.

The difference between these two numbers will be the probability for region III, which is $P(A \cap B) = 1.1 - 0.8 = 0.3$.

Since $P(A) \cdot P(B) = 0.6 \cdot 0.5 = 0.3$ and this is the same as $P(A \cap B)$, these are independent events.

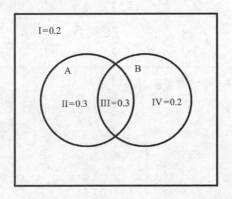

PART III

33. The zeros of a function p are the values of x for which $p(x) = 0$. To find the zeros of this function, solve the equation $0 = x^3 + x^2 - 4x - 4$. A polynomial equation with four terms like this can sometimes be factored with a technique called "factor by grouping." Factor an x^2 out of the first two terms, and factor -4 out of the last two terms. Luckily, there ends up being an $(x + 1)$ in each term that can then be further factored out:

$$0 = x^3 + x^2 - 4x - 4$$
$$0 = x^2(x+1) - 4(x+1)$$
$$0 = (x^2 - 4)(x+1)$$
$$0 = (x-2)(x+2)(x+1)$$

The solutions are $x = 2$, $x = -2$, and $x = -1$.

For the graph, the zeros of the function are related to the x-intercepts. The x-intercepts of the graph of p are $(2, 0)$, $(-2, 0)$, and $(-1, 0)$. Another useful point on the graph is the y-intercept, which you get by finding $p(0)$:

$$p(0) = 0^3 + 0^2 - 4(0) - 4 = -4$$

So the y-intercept is $(0, -4)$.

Plot the x-intercepts and the y-intercept on the graph. Connect them with a sideways "S" shape. You can use the graphing calculator to get an idea of what the graph should look like before you draw it.

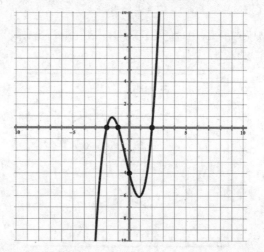

For the TI-84:   For the TI-Nspire:

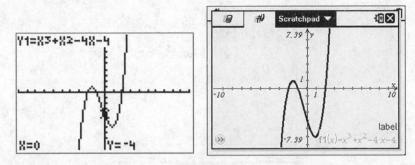

34. The formula for half-life is shown:

$$\text{amount remaining} = \text{initial amount} \cdot \left(\frac{1}{2}\right)^{\frac{t}{\text{half-life}}}$$

Substitute the given information into the formula:

$$7 = 20 \cdot \left(\frac{1}{2}\right)^{\frac{t}{8.02}}$$

Solving this equation for t will require you to use logarithms:

$$7 = 20 \cdot \left(\frac{1}{2}\right)^{\frac{t}{8.02}}$$

$$\frac{7}{20} = \frac{20 \cdot \left(\frac{1}{2}\right)^{\frac{t}{8.02}}}{20}$$

$$\frac{7}{20} = \left(\frac{1}{2}\right)^{\frac{t}{8.02}}$$

$$\log_{\frac{1}{2}}\left(\frac{7}{20}\right) = \frac{t}{8.02}$$

$$8.02 \cdot \log_{\frac{1}{2}}\left(\frac{7}{20}\right) = t$$

Enter this into either graphing calculator. The solution is $t \approx 12.14$, which rounds to 12 days.

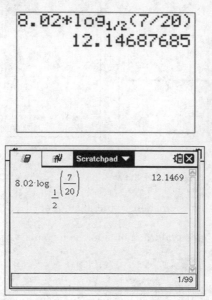

35. Isolate the radical expression by subtracting x from both sides of the equation and then squaring both sides of the equation:

$$\sqrt{2x-7} + x = 5$$
$$\underline{\phantom{\sqrt{2x-7}}-x = -x}$$
$$\sqrt{2x-7} = 5 - x$$
$$\left(\sqrt{2x-7}\right)^2 = \left(5-x\right)^2$$
$$2x - 7 = 25 - 10x + x^2$$

Solve the quadratic equation by first bringing every term over to one side. Then try to factor the resulting expression:

$$2x - 7 = 25 - 10x + x^2$$
$$\underline{-2x + 7 = -2x + 7}$$
$$0 = x^2 - 12x + 32$$
$$0 = (x-4)(x-8)$$

$$x - 4 = 0 \quad \text{or} \quad x - 8 = 0$$
$$x = 4 \quad \text{or} \quad x = 8$$

When solving a radical equation, there is a chance that an extraneous solution was formed in the step where both sides of the equation was squared. Both solutions must be substituted back into the original equation to see if this has happened.

Check $x = 4$:

$$\sqrt{2 \cdot 4 - 7} + 4 = 5$$
$$\sqrt{1} + 4 = 5$$
$$1 + 4 = 5$$
$$5 = 5$$

This is a solution to the equation.

Check $x = 8$:

$$\sqrt{2 \cdot 8 - 7} + 8 = 5$$
$$\sqrt{9} + 4 = 5$$
$$3 + 4 = 5$$
$$7 = 5$$

This is not a solution to the equation.

The $x = 8$ solution must be rejected, leaving $x = 4$ as the only solution.

36. a) There are several reasons why Ayva's hypothesis might be incorrect. One is that the sample size is very small, just 10 students per group. Another is that she cannot be sure that 10 minutes is enough time for the different drinks to impact reading speed.

b) In Ayva's initial experiment, the difference in the means, group 1 – group 2, was –1.4. The computer-generated simulations using the same 20 people split up into two groups randomly show that a difference of –1.4 is not so rare. A difference of 4 would be incredibly rare.

Since randomly splitting the 20 people into equal groups is not unlikely to produce a difference of means of –1.4, the difference is not statistically significant. So the difference is not an unusual occurrence.

PART IV

37. The formula for the amount of money at the end of x years is $f(x) = P\left(1 + \dfrac{r}{n}\right)^{nx}$, where P is the initial deposit, r is the interest rate, and n is the number of compoundings per year. For options A and B, this formula becomes:

$$A(x) = 5{,}000\left(1 + \frac{0.045}{1}\right)^{1x}$$

$$B(x) = 5{,}000\left(1 + \frac{0.046}{4}\right)^{4x}$$

Substitute $x = 6$ into both equations to find how much money each option will have after 6 years:

$$A(6) = 5{,}000\left(1 + \frac{0.045}{1}\right)^{1 \cdot 6} \approx 6{,}511.30$$

$$B(6) = 5{,}000\left(1 + \frac{0.046}{4}\right)^{4 \cdot 6} \approx 6{,}578.87$$

Find the difference of the two options:

$$6{,}578.87 - 6{,}511.30 = 67.57$$

Option B will earn \$67.57 more than option A.

To find the amount of time it takes for the money to double in option B, solve the equation $10{,}000 = 5{,}000 \left(1 + \dfrac{0.046}{4}\right)^{4x}$ using logarithms:

$$10{,}000 = 5{,}000 \left(1 + \frac{0.046}{4}\right)^{4x}$$

$$\frac{10{,}000}{5{,}000} = \frac{5{,}000 \left(1 + \dfrac{0.046}{4}\right)^{4x}}{5{,}000}$$

$$2 = \left(1 + \frac{0.046}{4}\right)^{4x}$$

$$2 = 1.0115^{4x}$$

$$\log_{1.0115} 2 = 4x$$

$$\frac{\log_{1.0115} 2}{4} = \frac{4x}{4}$$

$$\frac{\log_{1.0115} 2}{4} = x$$

Enter this equation into the calculator to solve it. The solution is $x \approx 15.15$, which rounds to the *nearest tenth of a year* to 15.2 years.

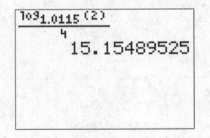

Topic	Question Numbers	Number of Points	Your Points	Your Percentage
1. Polynomial Expressions and Equations	5, 11, 15, 20, 21, 33	2 + 2 + 2 + 2 + 2 + 4 = 14		
2. Complex Numbers	1, 27	2 + 2 = 4		
3. Exponential Expressions and Equations	13, 22, 24, 34, 37	2 + 2 + 2 + 4 + 6 = 16		
4. Rational Expressions and Equations	17	2		
5. Radical Expressions and Equations	26, 35	2 + 4 = 16		
6. Trigonometric Expressions and Equations	10, 16, 25, 28	2 + 2 + 2 + 2 = 8		
7. Graphing	3, 6, 19	2 + 2 + 2 = 6		
8. Functions	14, 30, 31	2 + 2 + 2 = 6		
9. Systems of Equations	23	2		
10. Sequences and Series	8, 9, 18	2 + 2 + 2 = 6		
11. Probability	7, 32	2 + 2 = 4		
12. Statistics	2, 4, 12, 29, 36	2 + 2 + 2 + 2 + 4 = 12		

HOW TO CONVERT YOUR RAW SCORE TO YOUR ALGEBRA II REGENTS EXAMINATION SCORE

The accompanying conversion chart must be used to determine your final score on the August 2016 Regents Examination in Algebra II. To find your final exam score, locate in the column labeled "Raw Score" the total number of points you scored out of a possible 86 points. Since partial credit is allowed in Parts II, III, and IV of the test, you may need to approximate the credit you would receive for a solution that is not completely correct. Then locate in the adjacent column to the right the scale score that corresponds to your raw score. The scale score is your final Algebra II Regents Examination score.

Regents Examination in Algebra II—August 2016
Chart for Converting Total Test Raw Scores to Final
Examination Scores (Scaled Scores)

Raw Score	Scale Score	Performance Level	Raw Score	Scale Score	Performance Level	Raw Score	Scale Score	Performance Level
86	100	5	57	82	4	28	66	3
85	99	5	56	82	4	27	65	3
84	98	5	55	81	4	26	64	2
83	98	5	54	81	4	25	63	2
82	97	5	53	80	4	24	62	2
81	96	5	52	80	4	23	61	2
80	95	5	51	80	4	22	59	2
79	95	5	50	79	4	21	58	2
78	94	5	49	79	4	20	55	2
77	93	5	48	78	4	19	54	1
76	93	5	47	78	4	18	53	1
75	92	5	46	77	3	17	52	1
74	91	5	45	77	3	16	50	1
73	91	5	44	77	3	15	48	1
72	90	5	43	76	3	14	46	1
71	89	5	42	76	3	13	44	1
70	89	5	41	75	3	12	42	1
69	88	5	40	75	3	11	39	1
68	88	5	39	74	3	10	37	1
67	87	5	38	74	3	9	34	1
66	86	5	37	73	3	8	31	1
65	86	5	36	72	3	7	28	1
64	86	5	35	72	3	6	25	1
63	85	5	34	71	3	5	21	1
62	84	4	33	70	3	4	18	1
61	84	4	32	70	3	3	14	1
60	83	4	31	69	3	2	10	1
59	83	4	30	68	3	1	5	1
58	82	4	29	67	3	0	0	1

Examination
June 2017
Algebra II

HIGH SCHOOL MATH REFERENCE SHEET

Conversions

1 inch = 2.54 centimeters	1 cup = 8 fluid ounces
1 meter = 39.37 inches	1 pint = 2 cups
1 mile = 5280 feet	1 quart = 2 pints
1 mile = 1760 yards	1 gallon = 4 quarts
1 mile = 1.609 kilometers	1 gallon = 3.785 liters
	1 liter = 0.264 gallon
1 kilometer = 0.62 mile	1 liter = 1000 cubic centimeters
1 pound = 16 ounces	
1 pound = 0.454 kilogram	
1 kilogram = 2.2 pounds	
1 ton = 2000 pounds	

Formulas

Triangle	$A = \dfrac{1}{2}bh$
Parallelogram	$A = bh$
Circle	$A = \pi r^2$
Circle	$C = \pi d$ or $C = 2\pi r$

Formulas (continued)

General Prisms	$V = Bh$
Cylinder	$V = \pi r^2 h$
Sphere	$V = \dfrac{4}{3}\pi r^3$
Cone	$V = \dfrac{1}{3}\pi r^2 h$
Pyramid	$V = \dfrac{1}{3}Bh$
Pythagorean Theorem	$a^2 + b^2 = c^2$
Quadratic Formula	$x = \dfrac{-b \pm \sqrt{b^2 - 4ac}}{2a}$
Arithmetic Sequence	$a_n = a_1 + (n - 1)d$
Geometric Sequence	$a_n = a_1 r^{n-1}$
Geometric Series	$S_n = \dfrac{a_1 - a_1 r^n}{1 - r}$ where $r \neq 1$
Radians	$1 \text{ radian} = \dfrac{180}{\pi} \text{ degrees}$
Degrees	$1 \text{ degree} = \dfrac{\pi}{180} \text{ radians}$
Exponential Growth/Decay	$A = A_0 e^{k(t - t_0)} + B_0$

PART I

Answer all 24 questions in this part. Each correct answer will receive 2 credits. No partial credit will be allowed. For each statement or question, write in the space provided the numeral preceding the word or expression that best completes the statement or answers the question.   [48 credits]

1 The graph of the function $p(x)$ is sketched below.

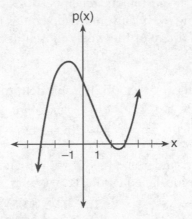

Which equation could represent $p(x)$?

(1) $p(x) = (x^2 - 9)(x - 2)$
(2) $p(x) = x^3 - 2x^2 + 9x + 18$
(3) $p(x) = (x^2 + 9)(x - 2)$
(4) $p(x) = x^3 - 2x^2 - 9x - 18$ 1 _____

2 What is the solution to $8(2^{x + 3}) = 48$?

(1) $x = \dfrac{\ln 6}{\ln 2} - 3$ (3) $x = \dfrac{\ln 48}{\ln 16} - 3$

(2) $x = 0$ (4) $x = \ln 4 - 3$ 2 _____

3 Cheap and Fast gas station is conducting a consumer satisfaction survey. Which method of collecting data would most likely lead to a biased sample?

   (1) interviewing every 5th customer to come into the station
   (2) interviewing customers chosen at random by a computer at the checkout
   (3) interviewing customers who call an 800 number posted on the customers' receipts
   (4) interviewing every customer who comes into the station on a day of the week chosen at random out of a hat   3 _____

4 The expression $6xi^3(-4xi + 5)$ is equivalent to

   (1) $2x - 5i$ (3) $-24x^2 + 30x - i$
   (2) $-24x^2 - 30xi$ (4) $26x - 24x^2i - 5i$ 4 _____

5 If $f(x) = 3|x| - 1$ and $g(x) = 0.03x^3 - x + 1$, an approximate solution for the equation $f(x) = g(x)$ is

   (1) 1.96   (3) $(-0.99, 1.96)$
   (2) 11.29   (4) $(11.29, 32.87)$ 5 _____

6 Given the parent function $p(x) = \cos x$, which phrase best describes the transformation used to obtain the graph of $g(x) = \cos(x + a) - b$, if a and b are positive constants?

   (1) right a units, up b units
   (2) right a units, down b units
   (3) left a units, up b units
   (4) left a units, down b units   6 _____

7 The solution to the equation $4x^2 + 98 = 0$ is

(1) ± 7

(3) $\pm \dfrac{7\sqrt{2}}{2}$

(2) $\pm 7i$

(4) $\pm \dfrac{7i\sqrt{2}}{2}$

7____

8 Which equation is represented by the graph shown below?

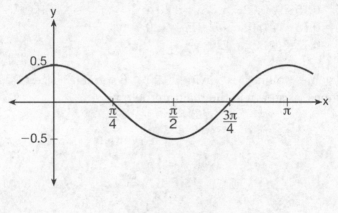

(1) $y = \dfrac{1}{2} \cos 2x$

(3) $y = \dfrac{1}{2} \cos x$

(2) $y = \cos x$

(4) $y = 2 \cos \dfrac{1}{2} x$

8____

9 A manufacturing company has developed a cost model, $C(x) = 0.15x^3 + 0.01x^2 + 2x + 120$, where x is the number of items sold, in thousands. The sales price can be modeled by $S(x) = 30 - 0.01x$. Therefore, revenue is modeled by $R(x) = x \cdot S(x)$.

The company's profit, $P(x) = R(x) - C(x)$, could be modeled by

(1) $0.15x^3 + 0.02x^2 - 28x + 120$
(2) $-0.15x^3 - 0.02x^2 + 28x - 120$
(3) $-0.15x^3 + 0.01x^2 - 2.01x - 120$
(4) $-0.15x^3 + 32x + 120$

9 _____

10 A game spinner is divided into 6 equally sized regions, as shown in the diagram below.

For Miles to win, the spinner must land on the number 6. After spinning the spinner 10 times, and losing all 10 times, Miles complained that the spinner is unfair. At home, his dad ran 100 simulations of spinning the spinner 10 times, assuming the probability of winning each spin is $\frac{1}{6}$. The output of the simulation is shown in the following diagram.

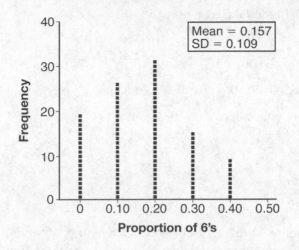

Which explanation is appropriate for Miles and his dad to make?

(1) The spinner was likely unfair, since the number 6 failed to occur in about 20% of the simulations.

(2) The spinner was likely unfair, since the spinner should have landed on the number 6 by the sixth spin.

(3) The spinner was likely not unfair, since the number 6 failed to occur in about 20% of the simulations.

(4) The spinner was likely not unfair, since in the output the player wins once or twice in the majority of the simulations.

10 _____

11 Which binomial is a factor of $x^4 - 4x^2 - 4x + 8$?

(1) $x - 2$ (3) $x - 4$

(2) $x + 2$ (4) $x + 4$ 11 _____

12 Given that $\sin^2 \theta + \cos^2 \theta = 1$ and $\sin \theta = -\dfrac{\sqrt{2}}{5}$, what is a possible value of $\cos \theta$?

(1) $\dfrac{5 + \sqrt{2}}{5}$ (3) $\dfrac{3\sqrt{3}}{5}$

(2) $\dfrac{\sqrt{23}}{5}$ (4) $\dfrac{\sqrt{35}}{5}$ 12 _____

13 A student studying public policy created a model for the population of Detroit, where the population decreased 25% over a decade. He used the model $P = 714(0.75)^d$, where P is the population, in thousands, d decades after 2010. Another student, Suzanne, wants to use a model that would predict the population after y years. Suzanne's model is best represented by

(1) $P = 714(0.6500)^y$ (3) $P = 714(0.9716)^y$

(2) $P = 714(0.8500)^y$ (4) $P = 714(0.9750)^y$ 13 _____

14 The probability that Gary and Jane have a child with blue eyes is 0.25, and the probability that they have a child with blond hair is 0.5. The probability that they have a child with both blue eyes and blond hair is 0.125. Given this information, the events blue eyes and blond hair are

         I: dependent
        II: independent
       III: mutually exclusive

(1) I, only   (3) I and III

(2) II, only   (4) II and III   14 _____

15 Based on climate data that have been collected in Bar Harbor, Maine, the average monthly temperature, in degrees F, can be modeled by the equation $B(x) = 23.914\sin(0.508x - 2.116) + 55.300$. The same governmental agency collected average monthly temperature data for Phoenix, Arizona, and found the temperatures could be modeled by the equation $P(x) = 20.238\sin(0.525x - 2.148) + 86.729$.

Which statement can *not* be concluded based on the average monthly temperature models x months after starting data collection?

(1) The average monthly temperature variation is more in Bar Harbor than in Phoenix.

(2) The midline average monthly temperature for Bar Harbor is lower than the midline temperature for Phoenix.

(3) The maximum average monthly temperature for Bar Harbor is 79°F, to the nearest degree.

(4) The minimum average monthly temperature for Phoenix is 20°F, to the nearest degree.   15 _____

16 For $x \neq 0$, which expressions are equivalent to one divided by the sixth root of x?

$$\text{I. } \frac{\sqrt[6]{x}}{\sqrt[3]{x}} \qquad \text{II. } \frac{x^{\frac{1}{6}}}{x^{\frac{1}{3}}} \qquad \text{III. } x^{-\frac{1}{6}}$$

(1) I and II, only   (3) II and III, only

(2) I and III, only   (4) I, II, and III   16 _____

17 A parabola has its focus at $(1, 2)$ and its directrix is $y = -2$. The equation of this parabola could be

(1) $y = 8(x + 1)^2$ (3) $y = 8(x - 1)^2$

(2) $y = \frac{1}{8}(x + 1)^2$ (4) $y = \frac{1}{8}(x - 1)^2$ 17 _____

18 The function $p(t) = 110e^{0.03922t}$ models the population of a city, in millions, t years after 2010. As of today, consider the following two statements:

    I.   The current population is 110 million.
    II. The population increases continuously by approximately 3.9% per year.

This model supports

(1) I, only   (3) both I and II

(2) II, only   (4) neither I nor II   18 _____

19 To solve $\dfrac{2x}{x-2} - \dfrac{11}{x} = \dfrac{8}{x^2 - 2x}$, Ren multiplied both sides by the least common denominator. Which statement is true?

(1) 2 is an extraneous solution.

(2) $\dfrac{7}{2}$ is an extraneous solution.

(3) 0 and 2 are extraneous solutions.

(4) This equation does not contain any extraneous solutions.   19 _____

20 Given $f(9) = -2$, which function can be used to generate the sequence $-8, -7.25, -6.5, -5.75, \ldots$?

(1) $f(n) = -8 + 0.75n$
(2) $f(n) = -8 - 0.75(n - 1)$
(3) $f(n) = -8.75 + 0.75n$
(4) $f(n) = -0.75 + 8(n - 1)$ 20 _____

21 The function $f(x) = 2^{-0.25x} \cdot \sin\left(\dfrac{\pi}{2} x\right)$ represents a

damped sound wave function. What is the average rate of change for this function on the interval $[-7, 7]$, to the *nearest hundredth*?

(1) -3.66 (3) -0.26

(2) -0.30 (4) 3.36 21 _____

22 Mallory wants to buy a new window air conditioning unit. The cost for the unit is \$329.99. If she plans to run the unit three months out of the year for an annual operating cost of \$108.78, which function models the cost per year over the lifetime of the unit, $C(n)$, in terms of the number of years, n, that she owns the air conditioner?

(1) $C(n) = 329.99 + 108.78n$

(2) $C(n) = 329.99 + 326.34n$

(3) $C(n) = \dfrac{329.99 + 108.78n}{n}$

(4) $C(n) = \dfrac{329.99 + 326.34n}{n}$ 22 _____

23 The expression $\dfrac{-3x^2 - 5x + 2}{x^3 + 2x^2}$ can be rewritten as

(1) $\dfrac{-3x - 3}{x^2 + 2x}$ (3) $-3x^{-1} + 1$

(2) $\dfrac{-3x - 1}{x^2}$ (4) $-3x^{-1} + x^{-2}$ 23 _____

24 Jasmine decides to put \$100 in a savings account each month. The account pays 3% annual interest, compounded monthly. How much money, S, will Jasmine have after one year?

(1) $S = 100(1.03)^{12}$

(2) $S = \dfrac{100 - 100(1.0025)^{12}}{1 - 1.0025}$

(3) $S = 100(1.0025)^{12}$

(4) $S = \dfrac{100 - 100(1.03)^{12}}{1 - 1.03}$

24 _____

PART II

Answer all 8 questions in this part. Each correct answer will receive 2 credits. Clearly indicate the necessary steps, including appropriate formula substitutions, diagrams, graphs, charts, etc. For all questions in this part, a correct numerical answer with no work shown will receive only 1 credit. [16 credits]

25 Given $r(x) = x^3 - 4x^2 + 4x - 6$, find the value of $r(2)$.

What does your answer tell you about $x - 2$ as a factor of $r(x)$? Explain.

26 The weight of a bag of pears at the local market averages 8 pounds with a standard deviation of 0.5 pound. The weights of all the bags of pears at the market closely follow a normal distribution. Determine what percentage of bags, to the *nearest integer*, weighed *less* than 8.25 pounds.

27 Over the set of integers, factor the expression $4x^3 - x^2 + 16x - 4$ completely.

28 The graph below represents the height above the ground, h, in inches, of a point on a triathlete's bike wheel during a training ride in terms of time, t, in seconds.

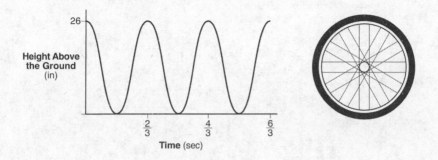

Identify the period of the graph and describe what the period represents in this context.

29 Graph $y = 400(.85)^{2x} - 6$ on the set of axes below.

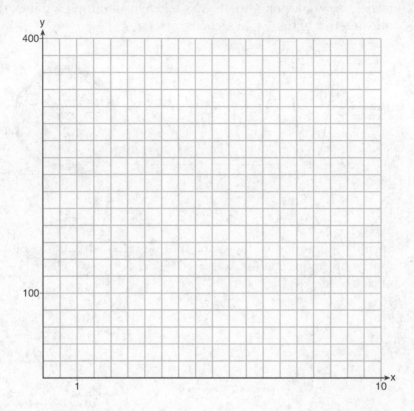

30 Solve algebraically for all values of x:

$$\sqrt{x-4} + x = 6$$

31 Write $\sqrt[3]{x} \cdot \sqrt{x}$ as a single term with a rational exponent.

32 Data collected about jogging from students with two older siblings are shown in the table below.

	Neither Sibling Jogs	One Sibling Jogs	Both Siblings Jog
Student Does Not Jog	1168	1823	1380
Student Jogs	188	416	400

Using these data, determine whether a student with two older siblings is more likely to jog if one sibling jogs or if both siblings jog. Justify your answer.

PART III

Answer all 4 questions in this part. Each correct answer will receive 4 credits. Clearly indicate the necessary steps, including appropriate formula substitutions, diagrams, graphs, charts, etc. For all questions in this part, a correct numerical answer with no work shown will receive only 1 credit. [16 credits]

33 Solve the following system of equations algebraically for all values of x, y, and z:

$$x + y + z = 1$$
$$2x + 4y + 6z = 2$$
$$-x + 3y - 5z = 11$$

34 Jim is looking to buy a vacation home for \$172,600 near his favorite southern beach. The formula to compute a mortgage payment, M, is $M = P \cdot \dfrac{r(1+r)^N}{(1+r)^N - 1}$ where P is the principal amount of the loan, r is the monthly interest rate, and N is the number of monthly payments. Jim's bank offers a monthly interest rate of 0.305% for a 15-year mortgage.

With no down payment, determine Jim's mortgage payment, rounded to the *nearest dollar*.

Algebraically determine and state the down payment, rounded to the *nearest dollar*, that Jim needs to make in order for his mortgage payment to be \$1100.

35 Graph $y = \log_2(x + 3) - 5$ on the set of axes below. Use an appropriate scale to include *both* intercepts.

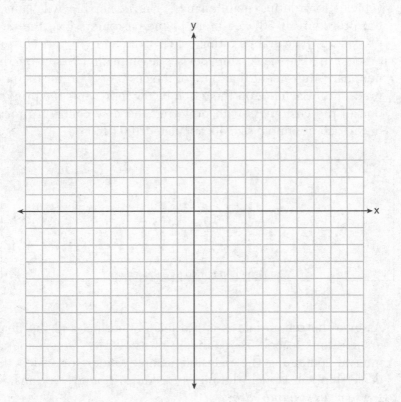

Describe the behavior of the given function as x approaches -3 and as x approaches positive infinity.

36 Charlie's Automotive Dealership is considering implementing a
new check-in procedure for customers who are bringing their
vehicles for routine maintenance. The dealership will launch
the procedure if 50% or more of the customers give the new
procedure a favorable rating when compared to the current
procedure. The dealership devises a simulation based on the
minimal requirement that 50% of the customers prefer the new
procedure. Each dot on the graph below represents the proportion of the customers who preferred the new check-in procedure, each of sample size 40, simulated 100 times.

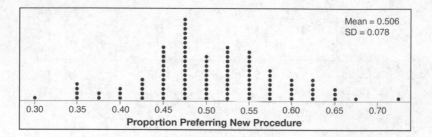

Assume the set of data is approximately normal and the dealership wants to be 95% confident of its results. Determine an
interval containing the plausible sample values for which the
dealership will launch the new procedure. Round your answer
to the *nearest hundredth*.

Forty customers are selected randomly to undergo the new
check-in procedure and the proportion of customers who prefer the new procedure is 32.5%. The dealership decides *not* to
implement the new check-in procedure based on the results of
the study. Use statistical evidence to explain this decision.

PART IV

Answer the question in this part. A correct answer will receive 6 credits. Clearly indicate the necessary steps, including appropriate formula substitutions, diagrams, graphs, charts, etc. A correct numerical answer with no work shown will receive only 1 credit. [6 credits]

37 A radioactive substance has a mass of 140 g at 3 p.m. and 100 g at 8 p.m. Write an equation in the form $A = A_0\left(\dfrac{1}{2}\right)^{\frac{t}{h}}$ that models this situation, where h is the constant representing the number of hours in the half-life, A_0 is the initial mass, and A is the mass t hours after 3 p.m.

Using this equation, solve for h, to the *nearest ten thousandth*.

Determine when the mass of the radioactive substance will be 40 g. Round your answer to the *nearest tenth of an hour*.

316

Answers
June 2017

Algebra II

Answer Key

PART I

1. (1)	**5.** (2)	**9.** (2)	**13.** (3)	**17.** (4)	**21.** (3)
2. (1)	**6.** (4)	**10.** (3)	**14.** (2)	**18.** (2)	**22.** (3)
3. (3)	**7.** (4)	**11.** (1)	**15.** (4)	**19.** (1)	**23.** (4)
4. (2)	**8.** (1)	**12.** (2)	**16.** (4)	**20.** (3)	**24.** (2)

PART II

25. -6, $(x - 2)$ is not a factor of $r(x)$

26. 69%

27. $(x^2 + 4)(4x - 1)$

28. Period is $\frac{2}{3}$. The wheel takes $\frac{2}{3}$ seconds to make one rotation.

29.

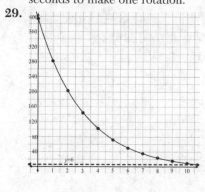

30. 5

31. $x^{\frac{5}{6}}$

32. More likely if both siblings jog

PART III

33. $(0, 2, -1)$

34. \$1247; \$20,407

35. y approaches negative infinity as x approaches -3. y approaches positive infinity as x approaches positive infinity.

36. The confidence interval is between 0.35 and 0.66; 0.325 is not between 0.35 and 0.66.

PART IV

37. $100 = 140\left(\frac{1}{2}\right)^{\frac{5}{h}}$; $h = 10.3002$; $x = 18.6$

In **Parts II–IV**, you are required to show how you arrived at your answers. For sample methods of solutions, see the *Answers Explained* section.

Answers Explained

PART I

1. When a graph has an x-intercept at $(a, 0)$, there is a factor of $(x - a)$. Since the graph has x-intercepts at -3, 2, and 3, the function could be:

$$p(x) = (x - (-3))(x - 2)(x - 3)$$
$$= (x + 3)(x - 2)(x - 3)$$

None of the answer choices looks exactly like this. Since

$x^2 - 9 = (x - 3)(x + 3)$, choice (1) is equivalent.

Another way to answer this is to graph the four answer choices on a graphing calculator to see which one most resembles the question.

When the function in choice (1) is graphed, it looks like the following.

For the TI-84:   For the TI-Nspire:

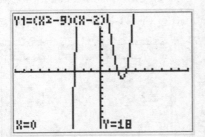 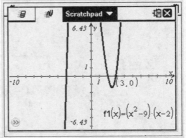

The correct choice is **(1)**.

2. To isolate the x, follow these four steps:

   Step 1: Divide both sides by 8.

$$\frac{8(2^{x+3})}{8} = \frac{48}{8}$$
$$2^{x+3} = 6$$

Step 2: Rearrange this exponential equation as a log equation. An equation of the form $a^b = c$ can be rewritten as $\log_a c = b$.

$$\log_2 6 = x + 3$$

Step 3: Subtract 3 from both sides of the equation.

$$\log_2 6 = x + 3$$
$$\underline{-3 = -3}$$
$$\log_2 6 - 3 = x$$

Step 4: Since this does not perfectly match any of the choices, you can use the change of base formula, $\log_b a = \dfrac{\log_c a}{\log_c b}$. Use $c = e$.

$$x = \frac{\log_e 6}{\log_e 2} - 3$$
$$= \frac{\ln 6}{\ln 2} - 3$$

Another way to answer this question is to test each of the four answer choices to see which makes the left side of the equation equal to 48.

For the TI-84:   For the TI-Nspire:

The correct choice is (**1**).

3. A self-selection bias is when the subjects get to choose whether or not they want to participate in a survey. This is not as random as picking every 5th customer, using a random number generator to decide who participates, or interviewing every customer on a random day. People who volunteer to be in a survey by calling a phone number may be the people who are either very satisfied or very unsatisfied. However, they may not represent the average customer.

The correct choice is (**3**).

4. First use the distributive property.

$$6xi^3(-4xi + 5) = 6xi^3 \cdot -4xi + 6xi^3 \cdot 5$$
$$= -24x^2i^4 + 30xi^3$$

Next, simplify the i-terms using the fact that $i^2 = -1$, $i^3 = -i$, and $i^4 = 1$. You could also use the graphing calculator to simplify the i-terms.

For the TI-84:   For the TI-Nspire:

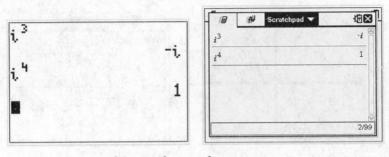

$$-24x^2i^4 + 30xi^3 = -24x^2(1) + 30x(-i)$$
$$= -24x^2 - 30xi$$

The correct choice is (2).

5. There are three values for x that satisfy the equation $f(x) = g(x)$. It is difficult to solve the equation $3|x| - 1 = 0.03x^3 - x + 1$ with algebra. So the best way to solve this is with the graphing calculator.

On the same set of axes, graph the following equations:

$$y = 3|x| - 1 \text{ and } y = 0.03x^3 - x + 1$$

Then locate the x-coordinates of the intersection points.

For the TI-84:   For the TI-Nspire:

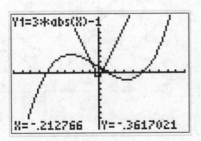 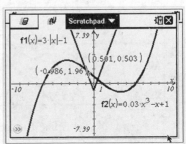

Based on this portion of the graph, two of the solutions are $x = -0.99$ and $x = 0.5$. It is tempting to answer choice (3) here. However, choice (3) is not correct because a solution to the equation is a value of x and not an ordered pair describing an intersection point.

If you zoom out on the graph, you will find another intersection point at $x = 11.29$.

For the TI-84:   For the TI-Nspire:

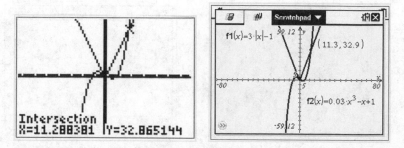

Since this is a multiple-choice question, the easiest way to solve it is to eliminate choices (3) and (4) first since they are ordered pairs instead of numbers. Then substitute the other choices into the two functions to see which results in the same output value for $f(x)$ and $g(x)$. Since $f(11.29) = 32.87$ and $g(11.29) = 32.87$, one solution is $x = 11.29$.

The correct choice is **(2)**.

6. If you pick values for a and b, like $a = 30$ and $b = 2$, and set your calculator to degree mode, you can use the graphing calculator to graph $p(x)$ and $g(x)$.

For the TI-84:   For the TI-Nspire:

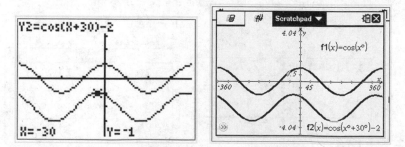

The graph of $g(x)$ is like the graph of $p(x)$ but shifted 30 to the left and 2 down.

In general, the graph of $p(x + a)$ will be the graph of $p(x)$ shifted a units to the left and the graph of $p(x) - b$ will be the graph of $p(x)$ shifted b units down.

The correct choice is (4).

7. Use algebra to isolate the x.

$$4x^2 = -98$$

$$\frac{4x^2}{4} = \frac{-98}{4}$$

$$x^2 = \frac{-98}{4}$$

$$\sqrt{x^2} = \pm\sqrt{-\frac{98}{4}}$$

$$x = \pm\frac{\sqrt{-98}}{\sqrt{4}}$$

$$= \pm\frac{\sqrt{2 \cdot 49 \cdot (-1)}}{2}$$

$$= \pm\frac{7i\sqrt{2}}{2}$$

The correct choice is (4).

8. A cosine curve with amplitude a and frequency b has the equation $y = a \cos(bx)$. In this curve, the amplitude is 0.5 since the highest point of the cosine curve is 0.5 above the middle of the curve. The period of the cosine curve is the difference between two consecutive high points. For this curve, the period is π because there is a high point at $(0, 0.5)$ and the next high point is at $(\pi, 0.5)$. The frequency can be calculated with the formula $b = \dfrac{2\pi}{\text{period}} = \dfrac{2\pi}{\pi} = 2$. Therefore, the equation is $y = 0.5 \cos(2x)$.

This could also be done by graphing the four choices on the graphing calculator to see which looks most like the graph. You would have to set your calculator to radians and use a window that matches the graph.

For the TI-84:   For the TI-Nspire:

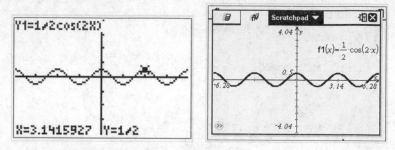

The correct choice is (**1**).

9. Subtract the polynomials to find the answer.

$$R(x) - C(x)$$
$$x \cdot S(x) - C(x)$$
$$x(30 - 0.01x) - (0.15x^3 + 0.01x^2 + 2x + 120)$$
$$30x - 0.01x^2 - 0.15x^3 - 0.01x^2 - 2x - 120$$
$$-0.15x^3 - 0.02x^2 + 28x - 120$$

The correct choice is (**2**).

10. If something is very unlikely, it will not happen very much in a computer simulation. Since 6 failed to occur about 20% of the time in a simulation of a fair spinner, it is common for a fair spinner not to come up with 6 in any of 10 tries. This spinner is likely to be a fair spinner (also described as "likely not unfair").

Had the simulation shown that a fair spinner would not land on 6 only 1% of the time, Miles would have had some evidence that the spinner he used was an unfair spinner.

The correct choice is (**3**).

11. The quickest way to check if something like $x - a$ is a factor of a polynomial is to use the factor theorem. To check if $(x - a)$ is a factor, substitute $x = a$ into the polynomial. If the polynomial evaluates to zero, then $(x - a)$ is a factor. If it evaluates to something other than zero, then $(x - a)$ is not a factor. To check if $(x + a)$ is a factor, substitute $x = -a$ into the polynomial. If the polynomial evaluates to zero, then $(x + a)$ is a factor. If it evaluates to something other than zero, then $(x + a)$ is not a factor. Test each choice.

Choice (1): Substitute $x = 2$ into the polynomial.

$$2^4 - 4 \cdot 2^2 - 4 \cdot 2 + 8 =$$
$$16 - 16 - 8 + 8 = 0$$

Since substituting 2 for x into the polynomial makes it evaluate to zero, $(x - 2)$ is a factor.

Choice (2): Substitute $x = -2$ into the polynomial.

$$(-2)^4 - 4 \cdot (-2)^2 - 4 \cdot (-2) + 8 =$$
$$16 - 16 + 8 + 8 = 16$$

Since substituting -2 for x into the polynomial makes it evaluate to something other than zero, $(x + 2)$ is a not a factor.

Choice (3): Substitute $x = 4$ into the polynomial.

$$(4)^4 - 4 \cdot (4)^2 - 4 \cdot (4) + 8 =$$
$$256 - 64 - 16 + 8 = 184$$

Since substituting 4 for x into the polynomial makes it evaluate to something other than zero, $(x - 4)$ is not a factor.

Choice (4): Substitute $x = -4$ into the polynomial

$$(-4)^4 - 4 \cdot (-4)^2 - 4 \cdot (-4) + 8 =$$
$$256 - 64 + 16 + 8 = 216$$

Since substituting -4 for x into the polynomial makes it evaluate to something other than zero, $(x + 4)$ is not a factor.

The correct choice is **(1)**.

12. Substitute the given value and solve for cos θ.

$$\sin^2\theta + \cos^2\theta = 1$$

$$\left(-\frac{\sqrt{2}}{5}\right)^2 + \cos^2\theta = 1$$

$$\frac{2}{25} + \cos^2\theta = 1$$

$$\underline{-\frac{2}{25} = -\frac{2}{25}}$$

$$\cos^2\theta = \frac{23}{25}$$

$$\sqrt{\cos^2\theta} = \pm\sqrt{\frac{23}{25}}$$

$$\cos\theta = \pm\frac{\sqrt{23}}{5}$$

The correct choice is **(2)**.

13. Since there are 10 years in a decade, $d = \frac{y}{10}$. Substitute for d in the equation.

$$P = 714(0.75)^{\frac{y}{10}}$$

To make this look more like the answer choices, rewrite $\frac{y}{10}$ as $\frac{1}{10} \cdot y$ and then use the rule of exponents, $x^{a \cdot b} = (x^a)^b$.

$$P = 714(0.75)^{\frac{1}{10} \cdot y}$$

$$= 714\left(0.75^{\frac{1}{10}}\right)^y$$

Then since $0.75^{\frac{1}{10}} \approx 0.9716$, the equation becomes $P = 714(0.9716)^y$.

Another way to solve this question is to pick a number for d such as $d = 5$ and calculate the value of $P = 714(0.75)^5 \approx 169$. Since 5 decades is the same as 50 years, check each of the answer choices with $y = 50$ to see which one gives the same answer.

For choice (3), $P = 714(0.9716)^{50} \approx 169$.

The correct choice is **(3)**.

14. Two events, A and B, are independent if $P(A \text{ and } B) = P(A) \cdot P(B)$. In this problem, event A is "child with blue eyes" and event B is "child with blond hair." It is given that $P(A) = 0.25$, $P(B) = 0.5$, and $P(A \text{ and } B) = 0.125$.

So $P(A) \cdot P(B) = 0.25 \cdot 0.5 = 0.125$ and $P(A \text{ and } B)$ also equals 0.125. When this happens, the events are independent.

Analyze each of the four statements.

Analyze I: When events are independent, they are not dependent.

Analyze II: These are independent events because

$$P(A) \cdot P(B) = P(A \text{ and } B)$$

Analyze III: Mutually exclusive occurs when $P(A \text{ and } B) = 0$, which is not the case in this problem since it is given that $P(A \text{ and } B) = 0.125$.

The correct choice is **(2)**.

15. The graphs of $B(x)$ and $P(x)$ can be used to analyze the answer choices.

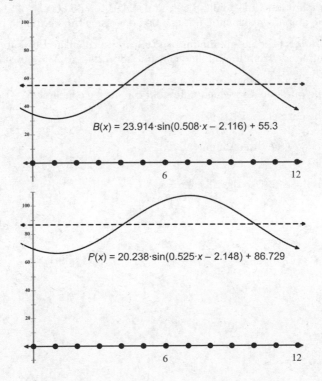

$B(x) = 23.914 \cdot \sin(0.508 \cdot x - 2.116) + 55.3$

$P(x) = 20.238 \cdot \sin(0.525 \cdot x - 2.148) + 86.729$

Choice (1): The city with a bigger difference between its high temperature and its low temperature has a higher average monthly temperature variation. Since $B(x)$ has a difference of around 48 between its high point and its low point while $P(x)$ has a difference of around 40 between its high point and its low point, $B(x)$ has a higher average monthly temperature variation.

Even without the graphs, it can be determined that the graph of $B(x)$ has a larger amplitude. The amplitude is the coefficient before the sine function. For $B(x)$, the amplitude is 23.914. For $P(x)$, the amplitude is 20.238.

Choice (2): The midline average monthly temperature for Bar Harbor is 55.3. The midline average monthly temperature for Phoenix is 86.729. So for Bar Harbor, the midline temperature is lower.

Without the graph, the midline can be determined by looking at the constant that is added to the sine term. For $B(x)$, this is 55.300. For $P(x)$, this is 86.729.

Choice (3): The maximum of $B(x)$ can be calculated by adding the amplitude to the midline constant, $23.914 + 55.300 = 79.214 \approx 79$. On the graphing calculator, this can be done using the maximum function.

Choice (4): The minimum of $P(x)$ can be calculated by subtracting the amplitude from the midline constant, $86.729 - 20.238 = 66.491$, which is not 20.

The question asks which statement can *not* be concluded.

The correct choice is **(4)**.

16. Analyze the three statements.

Analyze I: Use the rule $\sqrt[n]{x} = x^{\frac{1}{n}}$.

$$\frac{\sqrt[6]{x}}{\sqrt[3]{x}} = \frac{x^{\frac{1}{6}}}{x^{\frac{1}{3}}}$$

Then use the rule $\dfrac{x^a}{x^b} = x^{a-b}$.

$$\frac{x^{\frac{1}{6}}}{x^{\frac{1}{3}}} = x^{\frac{1}{6}-\frac{1}{3}}$$

$$= x^{\frac{1}{6}-\frac{2}{6}}$$

$$= x^{-\frac{1}{6}}$$

Finally use the rule $x^{-a} = \dfrac{1}{x^a}$.

$$x^{-\frac{1}{6}} = \frac{1}{x^{\frac{1}{6}}}$$

This is one divided by the sixth root of x. So statement I is an answer.

Analyze II: Use the rule $\dfrac{x^a}{x^b} = x^{a-b}$.

$$\frac{x^{\frac{1}{6}}}{x^{\frac{1}{3}}} = x^{\frac{1}{6}-\frac{1}{3}}$$

$$= x^{\frac{1}{6}-\frac{2}{6}}$$

$$= x^{-\frac{1}{6}}$$

This is identical to one of the steps when analyzing statement I. So statement II is an answer.

Analyze III: Use the rule $x^{-a} = \dfrac{1}{x^a}$.

$$x^{-\frac{1}{6}} = \frac{1}{x^{\frac{1}{6}}}$$

This is identical to one of the steps when analyzing statement I. So statement III is an answer.

So all three expressions are equivalent to one divided by the sixth root of x.

The correct choice is **(4)**.

17. A parabola with its focus at $(1, 2)$ and directrix at $y = -2$ looks like this.

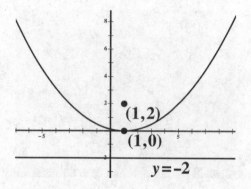

The vertex of a parabola is halfway between the vertex and the directrix. For this problem, the point halfway between $(1, 2)$ and $y = -2$ is $(1, 0)$.

The vertex form of the equation of a parabola is $y = a(x - h)^2 + k$, where the vertex is (h, k). Since the vertex is $(1, 0)$, the equation for this parabola will have the form $y = a(x - 1)^2 + 0 = y = a(x - 1)^2$. Only choices (3) and (4) have this form.

To calculate a, use the relationship $a = \dfrac{1}{4p}$, where p is half the distance between the focus and the directrix. Since the distance between the focus and the directrix is 4, p is 2. So $a = \dfrac{1}{4p} = \dfrac{1}{4 \cdot 2} = \dfrac{1}{8}$. The complete equation is $y = \dfrac{1}{8}(x - 1)^2$.

The correct choice is **(4)**.

18. In an exponential equation of the form $p(t) = Pe^{rt}$, the P is the initial value and the r is the annual growth rate compounded continuously. Since P is 110 in this problem, 110 million is the number of people at time $t = 0$, which is 2010. Statement I is incorrect since the current year is not 2010. Statement II is correct. If the word *continuously* was not included, statement II would not be correct since the annual growth rate is actually $e^{0.03922} \approx 1.04$.

   The correct choice is **(2)**.

19. To solve a rational equation, first factor all the denominators.

$$\frac{2x}{x-2} - \frac{11}{x} = \frac{8}{x(x-2)}$$

   The least common denominator is $x(x - 2)$. Multiply both sides of the equation by the least common denominator and simplify.

$$x(x-2)\left(\frac{2x}{x-2} - \frac{11}{x}\right) = x(x-2)\left(\frac{8}{x(x-2)}\right)$$

$$\frac{x(x-2)2x}{x-2} - \frac{x(x-2)11}{x} = \frac{x(x-2)8}{x(x-2)}$$

$$2x^2 - (x-2)11 = 8$$

$$2x^2 - (11x - 22) = 8$$

$$2x^2 - 11x + 22 = 8$$

$$\underline{\qquad -8 = -8 \qquad}$$

$$2x^2 - 11x + 14 = 0$$

$$(2x - 7)(x - 2) = 0$$

Solving for x seems to give two answers.

$$2x - 7 = 0$$

$$\underline{+7 = +7} \qquad\qquad x - 2 = 0$$

$$2x = 7 \qquad \text{and} \qquad \underline{+2 = +2}$$

$$\frac{2x}{2} = \frac{7}{2} \qquad\qquad x = 2$$

$$x = \frac{7}{2}$$

However, the $x = 2$ solution is extraneous. If you substitute it into the original equation, then two of the denominators evaluate to zero.

The correct choice is **(1)**.

20. The easiest way to solve this question is to test the four choices to see which of them evaluates to $f(1) = -8$ and $f(9) = -2$.

Choice (3):

$$f(1) = -8.75 + 0.75 \cdot 1$$
$$= -8.75 + 0.75$$
$$= -8$$

$$f(9) = -8.75 + 0.75 \cdot 9$$
$$= -8.75 + 6.75$$
$$= -2$$

None of the other choices have $f(1) = -8$ and $f(9) = -2$.

Another way to answer this question is to look at the sequence and notice that the first term is -8, and the common difference is 0.75, since each term is 0.75 greater than the previous term. The formula for the nth term of an arithmetic sequence given on the reference sheet is $a_n = a_1 + (n - 1)d$, where d is the common difference and a_1 is the first term.

For this problem, it becomes

$$a_n = -8 + (n - 1)0.75 = -8 + 0.75n - 0.75 = -8.75 + 0.75n$$

The fact is that $f(9) = -2$ is not really needed, although it could have been used to calculate the value of d with the equation $a_n = a_1 + (n - 1)d$.

$$-2 = -8 + (9 - 1)d$$
$$-2 = -8 + 8d$$
$$\underline{+8 = +8}$$
$$6 = 8d$$

$$\frac{6}{8} = \frac{8d}{8}$$

$$\frac{3}{4} = 0.75 = d$$

Since it is simpler to calculate d by just looking at the first two terms of the sequence, $f(9) = -2$ is not a necessary detail.

The correct choice is **(3)**.

21. The average rate of change of a function f on an interval $[a, b]$ can be calculated with the average rate of change formula.

$$\text{average rate of change} = \frac{f(b) - f(a)}{b - a}$$

For this problem, use the following equation:

$$\frac{f(b) - f(a)}{b - a} = \frac{f(7) - f(-7)}{7 - (-7)}$$

To calculate the terms in the numerator, put your calculator into radian mode.

$$f(7) = 2^{-0.25 \cdot 7} \cdot \sin\left(\frac{\pi}{2} \cdot 7\right)$$

$$= -0.2973$$

$$f(-7) = 2^{-0.25 \cdot (-7)} \cdot \sin\left(\frac{\pi}{2} \cdot (-7)\right)$$

$$= 3.36359$$

$$\frac{f(7) - f(-7)}{7 - (-7)} = \frac{-0.2973 - 3.36359}{7 - (-7)}$$

$$= \frac{-3.6609}{14}$$

$$\approx -0.26$$

The correct choice is **(3)**.

22. The cost of having the air conditioner for one year is 329.99 + 108.78. For two years, the cost is 329.99 + 108.78 · 2. For three years, it is 329.99 + 108.78 · 3. How many months in the year Mallory uses the air conditioner does not matter since \$108.78 is the cost per year, not the cost per month. In general, the cost for n years is 329.99 + 108.78n. The question, though, asks for an expression for the cost per year over the lifetime of the unit, not the cost during any particular year. Therefore, divide the expression by n to get $C(n)$.

$$C(n) = \frac{329.99 + 108.78n}{n}$$

The correct choice is **(3)**.

23. To simplify this expression, first factor the numerator and the denominator. To make factoring the numerator easier, start by factoring out the -1. The resulting quadratic polynomial will be simpler to factor.

$$\frac{-3x^2 - 5x + 2}{x^3 + 2x^2} = \frac{-(3x^2 + 5x - 2)}{x^2(x+2)}$$

$$= \frac{-(3x-1)(x+2)}{x^2(x+2)}$$

$$= \frac{-(3x-1)}{x^2}$$

$$= \frac{-3x+1}{x^2}$$

$$= \frac{-3x}{x^2} + \frac{1}{x^2}$$

$$= \frac{-3}{x} + \frac{1}{x^2}$$

$$= -3x^{-1} + x^{-2}$$

The correct choice is **(4)**.

24. The monthly interest rate is $\dfrac{0.03}{12} = 0.0025$. If you assume that the money is put into the bank on the first of every month starting on January 1 and that the final value will be calculated right after the twelfth deposit is made on December 1, the first \$100 will be worth $\$100(1.0025)^{11}$ since it will be in the bank for 11 months. The second \$100 will be worth $\$100(1.0025)^{10}$ on December 1. The third \$100 will be worth $\$100(1.0025)^9$ on December 1, and so on. The twelfth deposit, made on December 1, will still be worth \$100 on that day. There are a total of twelve \$100 deposits. At the end of the year, they will have a combined value of $100(1.0025)^{11} + 100(1.0025)^{10} + 100(1.0025)^9 + \cdots + 100$. This is a geometric series. The following shows how this can be rewritten:

$$100 + 100(1.0025)^1 + 100(1.0025)^2 + \cdots + 100(1.0025)^{11}$$

On the reference sheet is the formula for the sum of a geometric series, $S_n = \dfrac{a_1 - a_1 r^n}{1-r}$, where a_1 is the first term, r is the common ratio, and n is the number of terms.

For this problem, the formula becomes $S_{12} = \dfrac{100 - 100(1.0025)^{12}}{1 - 1.0025}$.

Even without having this formula, it is possible to eliminate two choices by noticing that the total value must be more than \$1,200 since that's how much would be in the account even with no interest. Choice (1) evaluates to only \$142.58. Choice (3) evaluates to only \$103.04.

The correct choice is **(2)**.

PART II

25. To find $r(2)$, substitute 2 for every x in the equation $r(x)$.

$$r(2) = 2^3 - 4 \cdot 2^2 + 4 \cdot 2 - 6$$
$$= 8 - 16 + 8 - 6$$
$$= -6$$

The factor theorem says that if $f(a) = 0$ then $(x - a)$ is a factor of $f(x)$. The theorem also says that if $f(a) \neq 0$, then $(x - a)$ is not a factor of $f(x)$. Since $r(2) \neq 0$, $(x - 2)$ is not a factor of $r(x)$.

26. This can be calculated by using the "normal cdf" function of the graphing calculator. This function determines the percent that can be expected between two numbers. When the question asks about the percent that can be expected below some number, make the lower bound something negative and very large since there is no way to make a negative infinity as the lower bound. Make the "lower" –9999, the "upper" 8.25, the "μ" 8, and the "σ" 0.5.

For the TI-84:

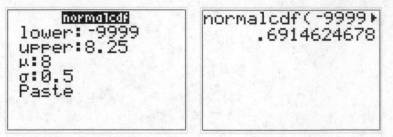

For the TI-Nspire:

When rounded to the nearest integer, the percentage of bags that weighed less than 8.25 pounds is 69%.

27. A cubic equation can sometimes be factored by grouping. Start by factoring out a common factor from the first two terms. Then factor out a common factor from the last two terms.

$$4x^3 - x^2 + 16x - 4$$
$$x^2(4x - 1) + 4(4x - 1)$$

If the two terms have a common factor (which they do in this case), factor that out too.

$$(4x - 1)(x^2 + 4)$$

These two factors cannot be factored anymore without using imaginary numbers, so this is considered completely factored over the set of integers.

28. The period of a cosine curve is the horizontal distance between one high point and the next high point. Since the first high point is at $(0, 26)$ and the second is at $(\frac{2}{3}, 26)$ the period of the graph is $\frac{2}{3}$.

In this context, the $\frac{2}{3}$ is the amount of time the wheel takes to make one rotation.

29. Make a chart with x-values from 0 to 10.

x	y
0	394
1	283
2	203
3	145
4	103
5	73
6	51
7	35
8	24
9	15
10	10

This makes an exponential graph. Graphs of exponential equations of the form $y = a \cdot b^x + c$ have a horizontal asymptote at $y = c$, so this graph has a horizontal asymptote at $y = 6$.

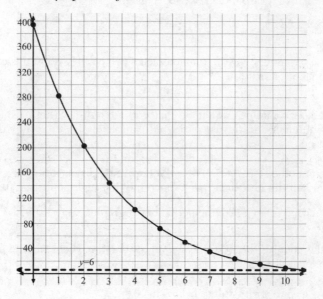

30. Isolate the radical by subtracting x from both sides of the equation. Then square both sides and solve the resulting quadratic equation.

$$\sqrt{x-4} + x = 6$$

$$\underline{-x = -x}$$

$$\sqrt{x-4} = 6 - x$$

$$\left(\sqrt{x-4}\right)^2 = (6-x)^2$$

$$x - 4 = 36 - 12x + x^2$$

$$\underline{-x = -x}$$

$$-4 = x^2 - 13x + 36$$

$$\underline{+4 = +4}$$

$$0 = x^2 - 13x + 40$$

$$0 = (x-5)(x-8)$$

$$x - 5 = 0 \quad \text{or} \quad x - 8 = 0$$

$$x = 5 \quad \text{or} \quad x = 0$$

Check both answers to see if either or both are extraneous solutions.

Check $x = 5$:

$$\sqrt{5-4}+5\overset{?}{=}6$$

$$\sqrt{1}+5\overset{?}{=}6$$

$$1+5=6$$

Check $x = 8$:

$$\sqrt{8-4}+8\overset{?}{=}6$$

$$\sqrt{4}+8\overset{?}{=}6$$

$$2+8\overset{?}{=}6$$

$$10\neq6$$

So 8 is an extraneous solution. The only solution is $x = 5$.

31. First use the rule $\sqrt[n]{x}=x^{\frac{1}{n}}$.

$$\sqrt[3]{x}\cdot\sqrt{x}=\sqrt[3]{x}\cdot\sqrt[2]{x}=x^{\frac{1}{3}}\cdot x^{\frac{1}{2}}$$

Then use the rule $x^a\cdot x^b=x^{a+b}$.

$$x^{\frac{1}{3}}\cdot x^{\frac{1}{2}}=x^{\frac{1}{3}+\frac{1}{2}}$$

$$=x^{\frac{2}{6}+\frac{3}{6}}$$

$$=x^{\frac{5}{6}}$$

32. There are a total of 416 + 1823 = 2239 students who have one sibling who jogs. Of those students, only 416 of them also jog. So the probability that a student with one sibling who jogs also jogs is $\dfrac{416}{2239} \approx 18.6\%$.

There are a total of 400 + 1380 = 1780 students who have two siblings who jog. Of those students, only 400 of them also jog. The probability that a student with two siblings who jog also jogs is $\dfrac{400}{1780} \approx 22.5\%$.

So it is more likely for a student who has two siblings who jog to jog than it is for a student who has only one sibling who jogs.

PART III

33. To solve a system of three equations with three unknowns, first find a way to combine two of the equations to eliminate one of the variables. Since the first equation has an x and the third equation has a $-x$, add the first and third equations. The result is a new equation that only has the variables y and z.

$$\begin{array}{r} x + y + z = 1 \\ -x + 3y - 5z = 11 \\ \hline 4y - 4z = 12 \end{array}$$

Next choose another pair of equations (either the first and second or the second and third). Eliminate the same variable, which is the x in this case. Eliminate the x from the second and third equations.

$$\begin{array}{l} 2x + 4y + 6z = 2 \\ -x + 3y - 5z = 11 \end{array}$$

$$\begin{array}{l} 2x + 4y + 6z = 2 \\ 2(-x + 3y - 5z) = 2(11) \end{array}$$

$$\begin{array}{r} 2x + 4y + 6z = 2 \\ + \;\; -2x + 6y - 10z = 22 \\ \hline 10y - 4z = 24 \end{array}$$

Now solve the system of two equations.

$$\begin{array}{l} 4y - 4z = 12 \\ 10y - 4z = 24 \end{array}$$

$$\begin{array}{l} 4y - 4z = 12 \\ -1(10y - 4z) = -1(24) \end{array}$$

$$\begin{array}{r} 4y - 4z = 12 \\ + \;\; -10y + 4z = -24 \\ \hline \dfrac{-6y}{-6} = \dfrac{-12}{-6} \end{array}$$

$$y = 2$$

Substitute $y = 2$ into one of the two equations that have just y and z in them and solve for z.

$$4(2) - 4z = 12$$
$$8 - 4z = 12$$
$$\underline{-8 = -8}$$
$$\frac{-4z}{-4} = \frac{4}{-4}$$
$$z = -1$$

Substitute $y = 2$ and $z = -1$ into one of the original three equations and solve for x.

$$x + 2 + (-1) = 1$$
$$x + 1 = 1$$
$$\underline{-1 = -1}$$
$$x = 0$$

The solution to the system is $x = 0$, $y = 2$, and $z = -1$, $(0, 2, -1)$.

34. Start by finding N. Since there are 12 months in a year, $12 \cdot 15 = 180$. Now substitute $P = 172{,}600$, $r = 0.00305$, and $N = 180$ into the equation, and solve for M.

$$M = 172{,}600 \cdot \frac{0.00305(1+0.00305)^{180}}{(1+0.00305)^{180}-1}$$
$$= 1247.49$$

Rounded to the nearest dollar, Jim's mortgage payment is $1247 with no down payment.

If Jim makes a down payment of \$$x$, the amount of his loan will be $172{,}600 - x$. Substitute $P = 172{,}600 - x$, $M = 1100$, $r = 0.00305$, and $N = 180$ into the equation, and solve for x.

$$1100 = (172{,}600 - x) \cdot \frac{0.00305(1+0.00305)^{180}}{(1+0.00305^{180})-1}$$

$$1100 = (172{,}600 - x) \cdot \frac{0.00305(1.00305)^{180}}{(1.00305^{180})-1}$$

$$1100 = (172{,}600 - x) \cdot 0.0072276558$$

$$\frac{1100}{0.0072276558} = \frac{(172{,}600 - x) \cdot 0.0072276558}{0.0072276558}$$

$$152{,}193.19 = 172{,}600 - x$$
$$\underline{-172{,}600 = -172{,}600}$$
$$-20{,}406.81 = -x$$
$$20{,}406.81 = x$$

Rounded to the nearest dollar, Jim needs to make a down payment of $20,407 in order for his mortgage payment to be $1100.

35. Before selecting the scale, determine the intercepts.

For the y-intercept, set $x = 0$ and solve for y.

$$y = \log_2(0 + 3) - 5$$
$$y = \log_2 3 - 5$$
$$y \approx -3.4$$

The y-intercept is $(0, -3.4)$.

For the x-intercept, set $y = 0$ and solve for x.

$$0 = \log_2(x + 3) - 5$$
$$\underline{+5 = +5}$$
$$5 = \log_2(x + 3)$$
$$2^5 = x + 3$$
$$32 = x + 3$$
$$\underline{-3 = -3}$$
$$29 = x$$

The x-intercept is $(29, 0)$.

By making each box 3 units, both intercepts will appear on the graph. Plot the two intercepts on the graph.

This can also be graphed on the graphing calculator. Set the window so that the x- and y-values go from -30 to $+30$.

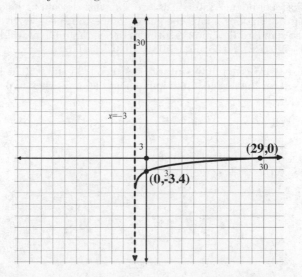

For the TI-84:   For the TI-Nspire:

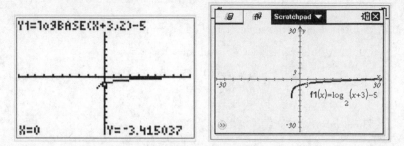

For x-values very close to -3 like -2.99999, the value of

$$\log_2(-2.9999 + 3) = \log_2(0.0001)$$

becomes a very large negative number. Since log is undefined for negative numbers, there is a vertical asymptote at $x = -3$. The behavior of the function as x approaches -3 is that the y-value approaches negative infinity.

Although it looks like there might be a horizontal asymptote, there is not one. For very large numbers like $x = 10,000,000$, the value of $\log_2(10,000,000 + 3) = \log_2(10,000,003)$ is well over 10. As the x-value increases, the y-value also increases. The range of a log graph is all real numbers. So as x approaches positive infinity, y also approaches positive infinity.

36. The 95% confidence interval is two standard deviations below and two standard deviations above the mean. Since the mean is given as 0.506 and the standard deviation is given as 0.078, the interval will be between $0.506 - 2(0.078) = 0.35$ and $0.506 + 2(0.078) = 0.66$.

Because 0.325 is not in the interval between 0.35 and 0.66, it is too low of a percent for the dealership to implement the new check-in procedure.

PART IV

37. Substitute $A_0 = 140$, $A = 100$, and $t = 5$ (the interval from 3 p.m. to 8 p.m.) into the equation.

$$100 = 140\left(\frac{1}{2}\right)^{\frac{5}{h}}$$

Solve this equation for h.

$$\frac{100}{140} = \frac{140\left(\frac{1}{2}\right)^{\frac{5}{h}}}{140}$$

$$\frac{5}{7} = \left(\frac{1}{2}\right)^{\frac{5}{h}}$$

$$\log_{\frac{1}{2}}\left(\frac{5}{7}\right) = \frac{5}{h}$$

$$0.4854 = \frac{5}{h}$$

$$0.4854h = 5$$

$$\frac{0.4854h}{0.4854} = \frac{5}{0.4854}$$

$$h = 10.3002$$

Substitute $A_0 = 140$, $A = 40$, and $h = 10.3002$ into the equation, and solve for t.

$$40 = 140\left(\frac{1}{2}\right)^{\frac{t}{10.3002}}$$

$$\frac{40}{140} = \frac{140\left(\frac{1}{2}\right)^{\frac{t}{10.3002}}}{140}$$

$$\frac{2}{7} = \left(\frac{1}{2}\right)^{\frac{t}{10.3002}}$$

$$\log_{\frac{1}{2}}\left(\frac{2}{7}\right) = \frac{t}{10.3002}$$

$$1.8073 = \frac{t}{10.3002}$$

$$18.6 = t$$

Topic	Question Numbers	Number of Points	Your Points	Your Percentage
1. Polynomial Expressions and Equations	9, 11, 17, 22, 23, 25, 27	2 + 2 + 2 + 2 + 2 + 2 + 2 = 14		
2. Complex Numbers	4, 7	2 + 2 = 4		
3. Exponential Expressions and Equations	2, 13, 16, 18, 24, 31, 34, 37	2 + 2 + 2 + 2 + 2 + 2 + 4 + 6 = 22		
4. Rational Expressions and Equations	19	2		
5. Radical Expressions and Equations	30	2		
6. Trigonometric Expressions and Equations	6, 8, 12, 15, 28	2 + 2 + 2 + 2 + 2 = 10		
7. Graphing	1, 5, 29, 35	2 + 2 + 2 + 4 = 10		
8. Functions	21	2		
9. Systems of Equations	33	4		
10. Sequences and Series	20	2		
11. Probability	14, 32	2 + 2 = 4		
12. Statistics	3, 10, 26, 36	2 + 2 + 2 + 4 = 10		

HOW TO CONVERT YOUR RAW SCORE TO YOUR ALGEBRA II REGENTS EXAMINATION SCORE

The accompanying conversion chart must be used to determine your final score on the June 2017 Regents Examination in Algebra II. To find your final exam score, locate in the column labeled "Raw Score" the total number of points you scored out of a possible 86 points. Since partial credit is allowed in Parts II, III, and IV of the test, you may need to approximate the credit you would receive for a solution that is not completely correct. Then locate in the adjacent column to the right the scale score that corresponds to your raw score. The scale score is your final Algebra II Regents Examination score.

Regents Examination in Algebra II—June 2017
Chart for Converting Total Test Raw Scores to Final
Examination Scores (Scaled Scores)

Raw Score	Scale Score	Performance Level	Raw Score	Scale Score	Performance Level	Raw Score	Scale Score	Performance Level
86	100	5	57	82	4	28	67	3
85	99	5	56	82	4	27	66	3
84	98	5	55	82	4	26	65	3
83	97	5	54	81	4	25	64	2
82	97	5	53	81	4	24	63	2
81	96	5	52	80	4	23	62	2
80	95	5	51	80	4	22	60	2
79	94	5	50	80	4	21	59	2
78	94	5	49	79	4	20	56	2
77	93	5	48	79	4	19	55	2
76	92	5	47	79	4	18	53	1
75	92	5	46	78	4	17	52	1
74	91	5	45	78	4	16	50	1
73	90	5	44	77	3	15	48	1
72	90	5	43	77	3	14	45	1
71	89	5	42	77	3	13	43	1
70	89	5	41	76	3	12	41	1
69	88	5	40	76	3	11	38	1
68	88	5	39	75	3	10	35	1
67	87	5	38	75	3	9	32	1
66	87	5	37	74	3	8	29	1
65	86	5	36	74	3	7	26	1
64	86	5	35	73	3	6	23	1
63	86	5	34	72	3	5	19	1
62	85	5	33	72	3	4	16	1
61	84	4	32	71	3	3	12	1
60	84	4	31	70	3	2	8	1
59	83	4	30	69	3	1	4	1
58	83	4	29	68	3	0	0	1

Examination August 2017

Algebra II

HIGH SCHOOL MATH REFERENCE SHEET

Conversions

1 inch = 2.54 centimeters

1 meter = 39.37 inches

1 mile = 5280 feet

1 mile = 1760 yards

1 mile = 1.609 kilometers

1 kilometer = 0.62 mile

1 pound = 16 ounces

1 pound = 0.454 kilogram

1 kilogram = 2.2 pounds

1 ton = 2000 pounds

1 cup = 8 fluid ounces

1 pint = 2 cups

1 quart = 2 pints

1 gallon = 4 quarts

1 gallon = 3.785 liters

1 liter = 0.264 gallon

1 liter = 1000 cubic centimeters

Formulas

Triangle	$A = \frac{1}{2}bh$
Parallelogram	$A = bh$
Circle	$A = \pi r^2$
Circle	$C = \pi d$ or $C = 2\pi r$

Formulas (continued)

General Prisms	$V = Bh$
Cylinder	$V = \pi r^2 h$
Sphere	$V = \frac{4}{3}\pi r^3$
Cone	$V = \frac{1}{3}\pi r^2 h$
Pyramid	$V = \frac{1}{3}Bh$
Pythagorean Theorem	$a^2 + b^2 = c^2$
Quadratic Formula	$x = \dfrac{-b \pm \sqrt{b^2 - 4ac}}{2a}$
Arithmetic Sequence	$a_n = a_1 + (n-1)d$
Geometric Sequence	$a_n = a_1 r^{n-1}$
Geometric Series	$S_n = \dfrac{a_1 - a_1 r^n}{1 - r}$ where $r \neq 1$
Radians	1 radian = $\dfrac{180}{\pi}$ degrees
Degrees	1 degree = $\dfrac{\pi}{180}$ radians
Exponential Growth/Decay	$A = A_0 e^{k(t - t_0)} + B_0$

PART I

Answer all 24 questions in this part. Each correct answer will receive 2 credits. No partial credit will be allowed. For each statement or question, write in the space provided the numeral preceding the word or expression that best completes the statement or answers the question. [48 credits]

1 The function $f(x) = \dfrac{x-3}{x^2 + 2x - 8}$ is undefined when

x equals

(1) 2 or –4   (3) 3, only

(2) 4 or –2   (4) 2, only   1_____

2 Which expression is equivalent to $(3k - 2i)^2$, where i is the imaginary unit?

(1) $9k^2 - 4$ (3) $9k^2 - 12ki - 4$

(2) $9k^2 + 4$ (4) $9k^2 - 12ki + 4$ 2_____

3 The roots of the equation $x^2 + 2x + 5 = 0$ are

(1) –3 and 1   (3) $-1 + 2i$ and $-1 - 2i$

(2) –1, only   (4) $-1 + 4i$ and $-1 - 4i$ 3_____

4 The solution set for the equation $\sqrt{x+14} - \sqrt{2x+5} = 1$ is

(1) {–6}   (3) {18}

(2) {2}   (4) {2, 22}   4_____

5 As x increases from 0 to $\dfrac{\pi}{2}$, the graph of the equation $y = 2 \tan x$ will

(1) increase from 0 to 2

(2) decrease from 0 to –2

(3) increase without limit

(4) decrease without limit   5_____

6 Which equation represents a parabola with the focus
at $(0, -1)$ and the directrix $y = 1$?

(1) $x^2 = -8y$ (3) $x^2 = 8y$

(2) $x^2 = -4y$ (4) $x^2 = 4y$ 6____

7 Which diagram represents an angle, α, measuring
$\dfrac{13\pi}{20}$ radians drawn in standard position, and its reference angle, θ?

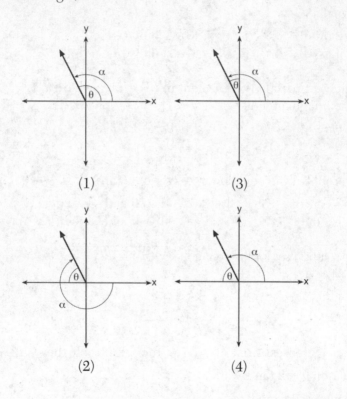

                              7____

8 What are the zeros of $P(m) = (m^2 - 4)(m^2 + 1)$?

(1) 2 and –2, only   (3) –4, i, and –i

(2) 2, –2, and –4   (4) 2, –2, i, and –i 8 _____

9 The value of a new car depreciates over time. Greg purchased a new car in June 2011. The value, V, of his car after t years can be modeled by the equation

$$\log_{0.8}\left(\frac{V}{17000}\right) = t.$$

What is the average decreasing rate of change per year of the value of the car from June 2012 to June 2014, to the *nearest ten dollars per year*?

(1) 1960   (3) 2450

(2) 2180   (4) 2770   9 _____

10 Iridium-192 is an isotope of iridium and has a half-life of 73.83 days. If a laboratory experiment begins with 100 grams of Iridium-192, the number of grams, A, of Iridium-192 present after t days would be

$$A = 100\left(\frac{1}{2}\right)^{\frac{t}{73.83}}.$$

Which equation approximates the amount of Iridium-192 present after t days?

(1) $A = 100\left(\dfrac{73.83}{2}\right)^t$ (3) $A = 100(0.990656)^t$

(2) $A = 100\left(\dfrac{1}{147.66}\right)^t$ (4) $A = 100(0.116381)^t$ 10 _____

11 The distribution of the diameters of ball bearings made under a given manufacturing process is normally distributed with a mean of 4 cm and a standard deviation of 0.2 cm. What proportion of the ball bearings will have a diameter less than 3.7 cm?

(1) 0.0668   (3) 0.8664

(2) 0.4332   (4) 0.9500   11_____

12 A polynomial equation of degree three, $p(x)$, is used to model the volume of a rectangular box. The graph of $p(x)$ has x intercepts at –2, 10, and 14. Which statements regarding $p(x)$ could be true?

A. The equation of $p(x) = (x - 2)(x + 10)(x + 14)$.
B. The equation of $p(x) = -(x + 2)(x - 10)(x - 14)$.
C. The maximum volume occurs when $x = 10$.
D. The maximum volume of the box is approximately 56.

(1) A and C (3) B and C

(2) A and D (4) B and D 12_____

13 Which expression is equivalent to $\dfrac{4x^3 + 9x - 5}{2x - 1}$, where $x \neq \dfrac{1}{2}$?

(1) $2x^2 + x + 5$ (3) $2x^2 - x + 5$

(2) $2x^2 + \dfrac{11}{2} + \dfrac{1}{2(2x-1)}$ (4) $2x^2 - x + 4 + \dfrac{1}{2x-1}$ 13_____

14 The inverse of the function $f(x) = \dfrac{x+1}{x-2}$ is

(1) $f^{-1}(x) = \dfrac{x+1}{x+2}$ (3) $f^{-1}(x) = \dfrac{x+1}{x-2}$

(2) $f^{-1}(x) = \dfrac{2x+1}{x-1}$ (4) $f^{-1}(x) = \dfrac{x-1}{x+1}$ 14_____

15 Which expression has been rewritten correctly to form a true statement?

(1) $(x + 2)^2 + 2(x + 2) - 8 = (x + 6)x$

(2) $x^4 + 4x^2 + 9x^2y^2 - 36y^2 = (x + 3y)^2(x - 2)^2$

(3) $x^3 + 3x^2 - 4xy^2 - 12y^2 = (x - 2y)(x + 3)^2$

(4) $(x^2 - 4)^2 - 5(x^2 - 4) - 6 = (x^2 - 7)(x^2 - 6)$

15_____

16 A study conducted in 2004 in New York City found that 212 out of 1334 participants had hypertension. Kim ran a simulation of 100 studies based on these data. The output of the simulation is shown in the diagram below.

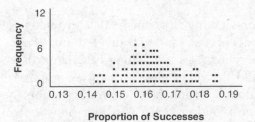

Proportion of Successes

At a 95% confidence level, the proportion of New York City residents with hypertension and the margin of error are closest to

(1) proportion $\approx .16$; margin of error $\approx .01$

(2) proportion $\approx .16$; margin of error $\approx .02$

(3) proportion $\approx .01$; margin of error $\approx .16$

(4) proportion $\approx .02$; margin of error $\approx .16$

16_____

17 Which scenario is best described as an observational study?

(1) For a class project, students in Health class ask every tenth student entering the school if they eat breakfast in the morning.

(2) A social researcher wants to learn whether or not there is a link between attendance and grades. She gathers data from 15 school districts.

(3) A researcher wants to learn whether or not there is a link between children's daily amount of physical activity and their overall energy level. During lunch at the local high school, she distributed a short questionnaire to students in the cafeteria.

(4) Sixty seniors taking a course in Advanced Algebra Concepts are randomly divided into two classes. One class uses a graphing calculator all the time, and the other class never uses graphing calculators. A guidance counselor wants to determine whether there is a link between graphing calculator use and students' final exam grades.

17 _____

18 Which sinusoid has the greatest amplitude?

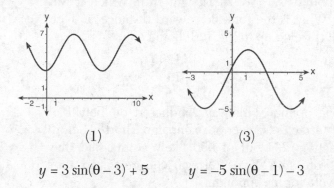

(1)   (3)

$y = 3\sin(\theta - 3) + 5$ $y = -5\sin(\theta - 1) - 3$

(2)   (4)   18 _____

19 Consider the system shown below.

$$2x - y = 4$$
$$(x + 3)^2 + y^2 = 8$$

The two solutions of the system can be described as

(1) both imaginary
(2) both irrational
(3) both rational
(4) one rational and one irrational   19 _____

20 Which binomial is *not* a factor of the expression $x^3 - 11x^2 + 16x + 84$?

(1) $x + 2$ (3) $x - 6$
(2) $x + 4$ (4) $x - 7$ 20 _____

21 A ball is dropped from a height of 32 feet. It bounces and rebounds 80% of the height from which it was falling. What is the total downward distance, in feet, the ball traveled up to the 12th bounce?

(1) 29   (3) 120

(2) 58   (4) 149   21 _____

22 A public opinion poll was conducted on behalf of Mayor Ortega's reelection campaign shortly before the election. 264 out of 550 likely voters said they would vote for Mayor Ortega; the rest said they would vote for his opponent.

Which statement is *least* appropriate to make, according to the results of the poll?

(1) There is a 48% chance that Mayor Ortega will win the election.

(2) The point estimate (p̂) of voters who will vote for Mayor Ortega is 48%.

(3) It is most likely that between 44% and 52% of voters will vote for Mayor Ortega.

(4) Due to the margin of error, an inference cannot be made regarding whether Mayor Ortega or his opponent is most likely to win the election.   22 _____

23 What does $\left(\dfrac{-54x^9}{y^4} \right)^{\frac{2}{3}}$ equal?

(1) $\dfrac{9ix^6 \sqrt[3]{4}}{y \sqrt[3]{y^2}}$ (3) $\dfrac{9x^6 \sqrt[3]{4}}{y \sqrt[3]{y}}$

(2) $\dfrac{9ix^6 \sqrt[3]{4}}{y^2 \sqrt[3]{y^2}}$ (4) $\dfrac{9x^6 \sqrt[3]{4}}{y^2 \sqrt[3]{y^2}}$ 23 _____

24 The Rickerts decided to set up an account for their daughter to pay for her college education. The day their daughter was born, they deposited $1000 in an account that pays 1.8% compounded annually. Beginning with her first birthday, they deposit an additional $750 into the account on each of her birthdays. Which expression correctly represents the amount of money in the account n years after their daughter was born?

(1) $a_n = 1000(1.018)^n + 750$

(2) $a_n = 1000(1.018)^n + 750n$

(3) $a_0 = 1000$

$a_n = a_{n-1}(1.018) + 750$

(4) $a_0 = 1000$

$a_n = a_{n-1}(1.018) + 750n$

24 _____

PART II

Answer all 8 questions in this part. Each correct answer will receive 2 credits. Clearly indicate the necessary steps, including appropriate formula substitutions, diagrams, graphs, charts, etc. For all questions in this part, a correct numerical answer with no work shown will receive only 1 credit. [16 credits]

25 Explain how $(-8)^{\frac{4}{3}}$ can be evaluated using properties of rational exponents to result in an integer answer.

26 A study was designed to test the effectiveness of a new drug. Half of the volunteers received the drug. The other half received a sugar pill. The probability of a volunteer receiving the drug and getting well was 40%. What is the probability of a volunteer getting well, given that the volunteer received the drug?

27 Verify the following Pythagorean identity for all values of x and y:

$$(x^2 + y^2)^2 = (x^2 - y^2)^2 + (2xy)^2$$

28 Mrs. Jones had hundreds of jelly beans in a bag that contained equal numbers of six different flavors. Her student randomly selected four jelly beans and they were all black licorice. Her student complained and said "What are the odds I got all of that kind?" Mrs. Jones replied, "simulate rolling a die 250 times and tell me if four black licorice jelly beans is unusual."

Explain how this simulation could be used to solve the problem.

29 While experimenting with her calculator, Candy creates the sequence 4, 9, 19, 39, 79, ...

Write a recursive formula for Candy's sequence.

Determine the eighth term in Candy's sequence.

30 In New York State, the minimum wage has grown exponentially. In 1966, the minimum wage was \$1.25 an hour and in 2015, it was \$8.75. Algebraically determine the rate of growth to the *nearest percent*.

31 Algebraically determine whether the function $j(x) = x^4 - 3x^2 - 4$ is odd, even, or neither.

32 On the axes below, sketch a possible function
$p(x) = (x - a)(x - b)(x + c)$, where a, b, and c are positive, $a > b$,
and $p(x)$ has a positive y-intercept of d. Label all intercepts.

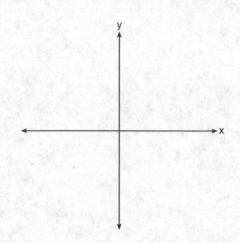

PART III

Answer all 4 questions in this part. Each correct answer will receive 4 credits. Clearly indicate the necessary steps, including appropriate formula substitutions, diagrams, graphs, charts, etc. For all questions in this part, a correct numerical answer with no work shown will receive only 1 credit. [16 credits]

33 Solve for all values of p: $\dfrac{3p}{p-5} - \dfrac{2}{p+3} = \dfrac{p}{p+3}$

34 Simon lost his library card and has an overdue library book. When the book was 5 days late, he owed \$2.25 to replace his library card and pay the fine for the overdue book. When the book was 21 days late, he owed \$6.25 to replace his library card and pay the fine for the overdue book.

Suppose the total amount Simon owes when the book is n days late can be determined by an arithmetic sequence. Determine a formula for a_n, the nth term of this sequence.

Use the formula to determine the amount of money, in dollars, Simon needs to pay when the book is 60 days late.

35 a) On the axes below, sketch *at least one* cycle of a sine curve with an amplitude of 2, a midline at $y = -\dfrac{3}{2}$, and a period of 2π.

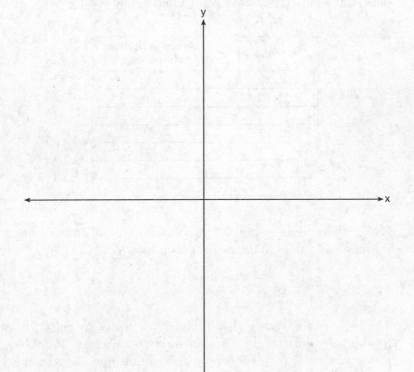

b) Explain any differences between a sketch of $y = 2\sin\left(x - \dfrac{\pi}{3}\right) - \dfrac{3}{2}$ and the sketch from part *a*.

36 Using a microscope, a researcher observed and recorded the number of bacteria spores on a large sample of uniformly sized pieces of meat kept at room temperature. A summary of the data she recorded is shown in the table below.

Hours (x)	Average Number of Spores (y)
0	4
0.5	10
1	15
2	60
3	260
4	1130
6	16,380

Using these data, write an exponential regression equation, rounding all values to the *nearest thousandth*.

The researcher knows that people are likely to suffer from food-borne illness if the number of spores exceeds 100. Using the exponential regression equation, determine the maximum amount of time, to the *nearest quarter hour*, that the meat can be kept at room temperature safely.

PART IV

Answer the question in this part. A correct answer will receive 6 credits. Clearly indicate the necessary steps, including appropriate formula substitutions, diagrams, graphs, charts, etc. A correct numerical answer with no work shown will receive only 1 credit. [6 credits]

37 The value of a certain small passenger car based on its use in years is modeled by $V(t) = 28482.698(0.684)^t$, where $V(t)$ is the value in dollars and t is the time in years. Zach had to take out a loan to purchase the small passenger car. The function $Z(t) = 22151.327(0.778)^t$, where $Z(t)$ is measured in dollars, and t is the time in years, models the unpaid amount of Zach's loan over time.

Graph $V(t)$ and $Z(t)$ over the interval $0 \le t \le 5$, on the set of axes below.

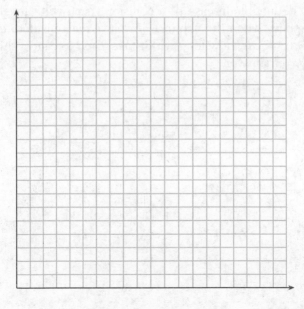

Question 37 is continued on the next page.

Question 37 continued.

State when $V(t) = Z(t)$, to *the nearest hundredth*, and interpret its meaning in the context of the problem.

Zach takes out an insurance policy that requires him to pay a $3000 deductible in case of a collision. Zach will cancel the collision policy when the value of his car equals his deductible. To the *nearest year*, how long will it take Zach to cancel this policy? Justify your answer.

Answers
August 2017
Algebra II

Answer Key

PART I

1. (1)	**5.** (3)	**9.** (3)	**13.** (1)	**17.** (2)	**21.** (4)
2. (3)	**6.** (2)	**10.** (3)	**14.** (2)	**18.** (4)	**22.** (1)
3. (3)	**7.** (4)	**11.** (1)	**15.** (1)	**19.** (1)	**23.** (4)
4. (2)	**8.** (4)	**12.** (4)	**16.** (2)	**20.** (2)	**24.** (3)

PART II

25. 16

26. 0.8

27. Both sides equal $x^4 + 2x^2y^2 + y^4$

28. See explanation

29. $a_1 = 4$
$a_n = 2a_{n-1} + 1$ for $n > 1$, 639

30. 4%

31. Even

32. See explanation for graph

PART III

33. $\{-5, -1\}$

34. $a_n = 1.25 + (n - 1) \cdot 0.25$, 16

35. See explanation for graph. The graph of the equation is like the original graph but shifted to the right.

36. $y = 4.168 \cdot 3.981^x$, 2.25 hours.

PART IV

37. See explanation for graph, 1.95 years, 6 years

In **Parts II–IV**, you are required to show how you arrived at your answers. For sample methods of solutions, see the *Answers Explained* section.

Answers Explained

PART I

1. A fraction is undefined when the denominator is 0. Since this function has a denominator, the values of x that make that denominator equal 0 are the ones that make the function undefined. To find these values, solve the equation:

$$x^2 + 2x - 8 = 0$$
$$(x + 4)(x - 2) = 0$$

$$\begin{array}{ccc} x + 4 = 0 & & x - 2 = 0 \\ \underline{-4 = -4} & \text{or} & \underline{+2 = +2} \\ x = -4 & & x = 2 \end{array}$$

The correct choice is (**1**).

2. Start by multiplying and combining like terms:

$$(3k - 2i)(3k - 2i) =$$
$$(3k)(3k) + (3k)(-2i) - (2i)(3k) + (-2i)(-2i) =$$
$$9k^2 - 6ki - 6ki + 4i^2 =$$
$$9k^2 - 12ki + 4i^2$$

Since i is the imaginary unit, $i = \sqrt{-1}$ and $i^2 = -1$. Replace i^2 with -1 in the expression:

$$9k^2 - 12ki + 4i^2 =$$
$$9k^2 - 12ki + 4(-1) =$$
$$9k^2 - 12ki - 4$$

The correct choice is (**3**).

3. Use the quadratic formula to find the values of x that satisfy this equation. Use $a = 1, b = 2,$ and $c = 5$:

$$x = \frac{-b \pm \sqrt{b^2 - 4ac}}{2a}$$

$$x = \frac{-2 \pm \sqrt{2^2 - 4 \cdot 1 \cdot 5}}{2 \cdot 1}$$

$$x = \frac{-2 \pm \sqrt{4 - 20}}{2} = \frac{-2 \pm \sqrt{-16}}{2} = \frac{-2 \pm \sqrt{16(-1)}}{2}$$

$$= \frac{-2 \pm \sqrt{16}\sqrt{-1}}{2} = \frac{-2 \pm 4i}{2} = -1 \pm 2i$$

The correct choice is **(3)**.

4. To solve this equation using algebra, first isolate one of the radicals and then square both sides of the equation:

$$\sqrt{x+14} - \sqrt{2x+5} = 1$$
$$\underline{+\sqrt{2x+5} = +\sqrt{2x+5}}$$
$$\sqrt{x+14} = 1 + \sqrt{2x+5}$$
$$\left(\sqrt{x+14}\right)^2 = \left(1 + \sqrt{2x+5}\right)^2$$
$$x+14 = \left(1 + \sqrt{2x+5}\right)\left(1 + \sqrt{2x+5}\right)$$
$$x+14 = 1 + 1\sqrt{2x+5} + 1\sqrt{2x+5} + 2x+5$$
$$x+14 = 1 + 2\sqrt{2x+5} + 2x+5$$
$$x+14 = 2x+6 + 2\sqrt{2x+5}$$

Now isolate the remaining radical, and square both sides of the equation:

$$x+14 = 2x+6 + 2\sqrt{2x+5}$$
$$\underline{-2x-6 = -2x-6}$$
$$-x+8 = 2\sqrt{2x+5}$$
$$(-x+8)^2 = \left(2\sqrt{2x+5}\right)^2$$
$$(-x+8)(-x+8) = \left(2\sqrt{2x+5}\right)^2$$
$$(-x)(-x) + (-x)8 + 8(-x) + 8 \cdot 8 = 2^2(2x+5)$$
$$x^2 - 16x + 64 = 4(2x+5)$$
$$x^2 - 16x + 64 = 8x + 20$$

Simplify and solve the quadratic equation by moving all terms to one side of the equation and factoring:

$$x^2 - 16x + 64 = 8x + 20$$
$$\underline{-8x - 20 = -8x - 20}$$
$$x^2 - 24x + 44 = 0$$
$$(x - 22)(x - 2) = 0$$

$$
\begin{array}{ccc}
x - 22 = 0 & & x - 2 = 0 \\
\underline{+22 = +22} & \text{or} & \underline{+2 = +2} \\
x = 22 & & x = 2
\end{array}
$$

At this point, it seems like the answer is choice (4). When solving radical equations, though, there are often "extraneous" solutions. This means some solutions to the quadratic equation are not solutions to the original equation. To see if either of these solutions need to be rejected, test both to see if they make the original equation true.

Test $x = 2$:

$$\sqrt{2 + 14} - \sqrt{2 \cdot 2 + 5} \overset{?}{=} 1$$
$$\sqrt{16} - \sqrt{9} \overset{?}{=} 1$$
$$4 - 3 \overset{?}{=} 1$$
$$1 = 1$$
$$\checkmark$$

Test $x = 22$:

$$\sqrt{22 + 14} - \sqrt{2 \cdot 22 + 5} \overset{?}{=} 1$$
$$\sqrt{36} - \sqrt{49} \overset{?}{=} 1$$
$$6 - 7 \overset{?}{=} 1$$
$$-1 \neq 1$$

Since $x = 22$ is not a solution to the original equation, the complete solution set is $\{2\}$.

Solving this question with algebra is very time consuming and difficult to do without making some careless error. If this was a Part II, III, or IV question, you would have to do it that way. Since this is a multiple-choice question, you could just test the four values used in the answer choices to see which of them make the equation true. Since only 2 makes the equation true, it must be the solution set.

The correct choice is (**2**).

5. The graph of $y = 2 \tan x$ looks like this:

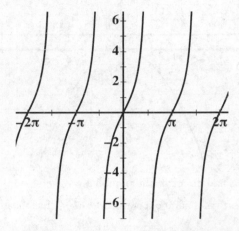

This graph can be produced on the graphing calculator if you put the calculator into radian mode.

For the TI-84:

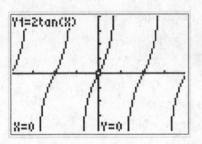

For the TI-Nspire:

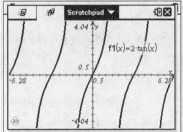

There is a vertical asymptote at $x = \dfrac{\pi}{2}$, so the curve gets closer and closer to the asymptote without ever reaching it. The closer the curve gets, the higher the y-coordinate is. So we can say that the graph increases without limit.

The correct choice is (**3**).

6. If you are given the focus and directrix of a parabola, the vertex of the parabola will be halfway between the focus and the directrix. The parabola curves around the focus away from the directrix.

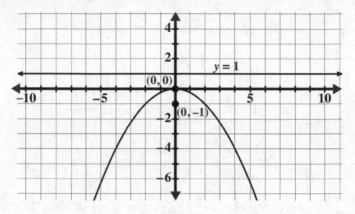

For this focus and directrix, the point halfway between them is $(0, 0)$.

The vertex form of a parabola with vertex (h, k) has the equation:

$$y = \pm a(x - h)^2 + k$$

For this example, h and k are both 0. When the focus is above the directrix, a will be positive. In this case, the focus is below the directrix, so a will be negative. The value of a can be calculated with the formula $a = \dfrac{1}{4p}$, where p is the distance between the vertex and the focus.

Since the vertex is 1 unit from the focus, the value of p is 1, so:

$$a = \frac{1}{4p} = \frac{1}{4 \cdot 1} = \frac{1}{4}$$

The equation of the parabola is $y = -\dfrac{1}{4}(x - 0)^2 + 0 = -\dfrac{1}{4}x^2$. After multiplying both sides of the equation by -4, it becomes $-4y = x^2$.

The correct choice is (**2**).

7. An angle in standard position has the positive x-axis as its initial ray and another ray rotated counterclockwise as its terminal ray. Choice (2) is not in standard position, so it can be eliminated. In degrees, angle α can be calculated by multiplying the radian measure by $\dfrac{180°}{\pi}$:

$$\alpha = \frac{13\pi}{20} \times \frac{180°}{\pi} = 117°$$

The reference angle θ is always an acute angle, so choice (1) can also be eliminated.

When an angle terminates in quadrant II, the reference angle is the acute angle formed by the negative x-axis and the terminal ray. This is represented in choice (4).

The measure of the quadrant II reference angle can also be found by subtracting the original angle from 180°:

$$\begin{aligned} \theta &= 180° - \alpha \\ &= 180° - 117° \\ &= 63° \end{aligned}$$

The measure of θ in choice (3) appears to be less than 63°.

The correct choice is (**4**).

8. The zeros of a function are the input values that will make the function evaluate to zero. To find the zeros of P, solve the equation:

$$\begin{aligned} 0 &= (m^2 - 4)(m^2 + 1) \\ 0 &= (m - 2)(m + 2)(m^2 + 1) \end{aligned}$$

If any of these factors is equal to zero, the product will be zero:

$$
\begin{array}{ccccc}
m - 2 = 0 & & m + 2 = 0 & & m^2 + 1 = 0 \\
\underline{+2 = +2} & \text{or} & \underline{-2 = -2} & \text{or} & \underline{-1 = -1} \\
m = 2 & & m = -2 & & m^2 = -1 \\
& & & & \sqrt{m^2} = \pm\sqrt{-1} \\
& & & & m = \pm i
\end{array}
$$

The correct choice is (**4**).

9. The average rate of change can be calculated using the following formula:

$$\text{Average Rate of Change} = \frac{\text{Value in 2014} - \text{Value in 2012}}{2014 - 2012}$$

Calculate the values in the numerator using the property that $\log_b a = c$ can be rewritten as $b^c = a$.

Calculate the value in 2012 ($t = 1$):

$$\log_{0.8}\left(\frac{V}{17000}\right) = 1$$

$$0.8^1 = \frac{V}{17000}$$

$$17000 \cdot 0.8^1 = 17000 \cdot \frac{V}{17000}$$

$$13600 = V$$

Calculate the value in 2014 ($t = 3$):

$$\log_{0.8}\left(\frac{V}{17000}\right) = 3$$

$$0.8^3 = \frac{V}{17000}$$

$$17000 \cdot 0.8^3 = 17000 \cdot \frac{V}{17000}$$

$$8704 = V$$

$$\text{Average Rate of Change} = \frac{\text{Value in 2014} - \text{Value in 2012}}{2014 - 2012}$$

$$= \frac{8704 - 13600}{2014 - 2012}$$

$$= \frac{-4896}{2}$$

$$\approx -2450$$

Since the question asks for the average decreasing rate of change to the nearest ten dollars, the answer is +2450 instead of −2450.

The correct choice is (3).

10. A property of exponents is that $x^{ab} = (x^a)^b$. This can be used to write the given equation as an equivalent equation:

$$A = 100\left(\frac{1}{2}\right)^{\frac{t}{73.83}}$$

$$= 100\left(\frac{1}{2}\right)^{\frac{1}{73.83} \cdot t}$$

$$= 100\left(\frac{1}{2}^{\frac{1}{73.83}}\right)^t$$

$$\approx 100(0.990656)^t$$

The easiest way to solve this question, however, would just be to pick a number like $t = 10$ and use the given equation to calculate the value of A:

$$A = 100\left(\frac{1}{2}\right)^{\frac{10}{73.83}} \approx 91$$

Then check each of the answer choices to see which of them also produces a value of A equal to 91 for $t = 10$.

Test choice (3):

$$A = 100(0.990656)^{10} \approx 91$$

The correct choice is **(3)**.

11. Use the normal cdf feature of the graphing calculator. The upper bound is 3.7, the standard deviation σ is 0.2, and the mean μ is 4. For the lower bound (since there is no lower bound), use a large negative number like −9999.

For the TI-84:

Press [2ND] [VARS] [2] to access the normal cdf function.

For the TI-Nspire:

Press [menu] [5] [5] [2] to access the Normal Cdf function.

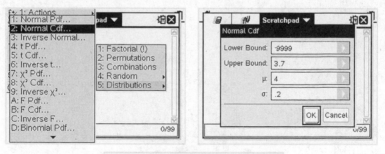

0.0668 is the percent of ball bearings with a diameter less than 3.7 cm in this normal distribution.

The correct choice is **(1)**.

12. When the graph of a polynomial function has an x-intercept at a, the function has a factor of $(x - a)$. This function will have as three of its factors $(x - (-2)) = (x + 2)$, $(x - 10)$, and $(x - 14)$.

Statement A does not have these factors, so it cannot be true.

Statement B does have the correct factors and also an additional factor of -1 because of the negative sign in front of the factors. Therefore this statement can be true.

At $x = 10$, the value of the function is 0, which is not the maximum. Statement (C) cannot be true.

If you graph $p(x) = -(x + 2)(x - 10)(x - 14)$ and use the maximum feature of the graphing calculator, it will tell you that the maximum value will occur at $x = 12.14$ and the maximum volume will be approximately 56.

For the TI-84:

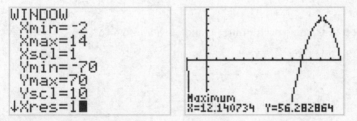

For the TI-Nspire:

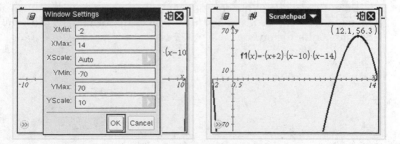

Both statements B and D could be true, while statements A and C could not be true.

The correct choice is **(4)**.

13. Divide $4x^3 + 9x - 5$ by $2x - 1$ using polynomial long division:

$$2x - 1 \overline{\smash{\big)}\ 4x^3 + 0x^2 + 9x - 5} \qquad \overset{\displaystyle 2x^2 + 1x + 5}{}$$

$$\begin{array}{r}
2x^2 + 1x + 5 \\
2x - 1 \overline{\smash{\big)}\ 4x^3 + 0x^2 + 9x - 5} \\
\underline{-(4x^3 - 2x^2)} \\
2x^2 + 9x - 5 \\
\underline{-(2x^2 - x)} \\
10x - 5 \\
\underline{-(10x - 5)} \\
0
\end{array}$$

The correct choice is (**1**).

14. Finding the inverse can be done in three steps:

Step 1: Change the $f(x)$ into a y.

$$y = \frac{x + 1}{x - 2}$$

Step 2: Change each x into a y and each y into an x.

$$x = \frac{y + 1}{y - 2}$$

Step 3: Solve for y in terms of x.

For this question, this last step is very involved because y occurs two times. Cross multiply, and then simplify. Move all the terms that have a y in them to one side of the equation and all other terms to the other side. Finally, factor out the y and divide both sides to isolate the y. This will be the answer for the inverse.

$$x = \frac{y+1}{y-2}$$

$$(y-2)x = y+1$$

$$xy - 2x = y+1$$

$$\underline{+2x = +2x}$$

$$xy = y+1+2x$$

$$\underline{-y = -y}$$

$$xy - y = 1+2x$$

$$y(x-1) = 1+2x$$

$$\frac{y(x-1)}{x-1} = \frac{1+2x}{x-1}$$

$$y = \frac{1+2x}{x-1}$$

$$y = \frac{2x+1}{x-1}$$

$$f^{-1}(x) = \frac{2x+1}{x-1}$$

The correct choice is **(2)**.

15. Test the choices until you find one that is true.

Choice (1):

$$(x+2)^2 + 2(x+2) - 8 \overset{?}{=} (x+6)x$$

$$(x+2)(x+2) + 2(x+2) - 8 \overset{?}{=} x^2 + 6x$$

$$x^2 + 2x + 2x + 4 + 2x + 4 - 8 \overset{?}{=} x^2 + 6x$$

$$x^2 + 6x = x^2 + 6x$$

$$\checkmark$$

The correct choice is **(1)**.

16. A plot of a sampling distribution of sample proportion has a proportion close to the median number of the data points. For this example, that is 0.16. The margin of error is the difference between the median and the highest and lowest numbers. Since the lowest numbers are approximately 0.02 away from the median and the highest numbers are also approximately 0.02 away from the median, the margin of error is approximately 0.02.

The correct choice is **(2)**.

17. In an observational study, the person doing the study must simply observe what the subjects are doing and not interfere or interact with the subjects in any way.

Choice (1) is a survey where the subjects answer questions rather than get observed.

Choice (2) is an observational study since the researcher is gathering the data without any interaction with the actual subjects.

Choice (3) is also a survey.

Choice (4) is called an experimental study since the researcher decides which of the subjects use calculators and which ones do not.

The correct choice is **(2)**.

18. The amplitude of a sine curve is the distance from the midline to the maximum of the curve. It can also be thought of as half the distance between the y-coordinates of the maximum point and the minimum point.

Choice (1): The amplitude is 2. The minimum has a y-coordinate of 3, and the maximum has a y-coordinate of 7. Half the distance between 3 and 7 is 2.

Choice (2): The amplitude is 4. The minimum has a y-coordinate of −5, and the maximum has a y-coordinate of 3. Half the distance between −5 and 3 is 4.

Choice (3): When the equation is given, the amplitude is the absolute value of the coefficient before the sine function. For this choice, the amplitude is 3.

Choice (4): The coefficient before the sine function is −5, so the amplitude is 5.

Of the four choices, 5 is the greatest amplitude.

The correct choice is **(4)**.

19. The best way to solve a system of a linear and quadratic equation is to isolate one of the variables in the linear equation, substitute that variable into the quadratic equation, and then solve.

Isolate the y-term in the linear equation:

$$2x - y = 4$$
$$\underline{-2x = -2x}$$
$$-y = -2x + 4$$
$$\frac{-y}{-1} = \frac{-2x + 4}{-1}$$
$$y = 2x - 4$$

Substitute into the quadratic equation:

$$(x+3)^2 + y^2 = 8$$
$$(x+3)^2 + (2x-4)^2 = 8$$
$$x^2 + 6x + 9 + 4x^2 - 16x + 16 = 8$$
$$5x^2 - 10x + 25 = 8$$
$$\underline{-8 = -8}$$
$$5x^2 - 10x + 17 = 0$$

Since the question does not ask for the solutions but for what type of solutions there will be, this can be completed by calculating the discriminant of the quadratic equation:

$$D = b^2 - 4ac$$

When the discriminant is negative, the equation has two imaginary solutions.

When the discriminant is zero, the equation has two equal rational solutions.

When the discriminant is a positive perfect square, the equation has two unequal rational solutions.

When the discriminant is positive but not a perfect square, the equation has two unequal irrational roots.

$$D = (-10)^2 - 4 \cdot 5 \cdot 17$$
$$= 100 - 340$$
$$= -240$$

Since the discriminant is negative, this equation—and also the original system of equations—has two imaginary solutions.

The correct choice is **(1)**.

20. The quickest way to check if an expression of the form $x + a$ (or $x - a$) is a factor of a polynomial is to use the factor theorem. The factor theorem says that if in a polynomial function $f(a) = 0$, then $(x - a)$ is a factor of $f(x)$. Test the four choices.

Choice (1): To check if $x + 2$ is a factor, substitute -2 for x into the polynomial:

$$(-2)^3 - 11(-2)^2 + 16(-2) + 84 = -8 - 44 - 32 + 84 = 0$$

Since $f(-2) = 0$, $x + 2$ is a factor.

Choice (2): To check if $x + 4$ is a factor, substitute -4 for x into the polynomial:

$$(-4)^3 - 11(-4)^2 + 16(-4) + 84 = -64 - 176 - 64 + 84 = -220 \neq 0$$

Since $f(-4) \neq 0$, $x + 4$ is *not* a factor.

Choice (3): To check if $x - 6$ is a factor, substitute $+6$ for x into the polynomial:

$$6^3 - 11 \cdot 6^2 + 16 \cdot 6 + 84 = 216 - 396 + 96 + 84 = 0$$

Since $f(6) = 0$, $x - 6$ is a factor.

Choice (4): To check if $x - 7$ is a factor, substitute $+7$ for x into the polynomial:

$$7^3 - 11 \cdot 7^2 + 16 \cdot 7 + 84 = 343 - 539 + 112 + 84 = 0$$

Since $f(7) = 0$, $x - 7$ is a factor.

Only $x + 4$ is not a factor.

The correct choice is **(2)**.

21. After the ball drops 32 feet, it bounces up to a height of $32 \times 0.8 = 25.6$ feet and then down again. By the second bounce, the ball has a total downward distance (which does not include the distance that it travels on the way back up) of $32 + 25.6$ feet, which is already about 58 feet. There are still 10 more bounces to go so choices (1) and (2) can be eliminated.

Without using a formula, the total downward distance can be calculated by making a chart with the downward distance for each of the twelve bounces and then adding them together. Each distance will be the previous distance multiplied by 0.8.

Bounce	Distance
1	32
2	25.6
3	20.48
4	16.38
5	13.11
6	10.49
7	8.39
8	6.71
9	5.37
10	4.29
11	3.44
12	2.75
Total	**149**

The "official" way to solve this question, however, is to recognize the pattern as a geometric series.

The downward distance for the first bounce is 32 feet.

The downward distance for the second bounce is $32 \cdot 0.8$ feet.

The downward distance for the third bounce is $32 \cdot 0.8^2$ feet.

The downward distance for the fourth bounce is $32 \cdot 0.8^3$ feet.

Notice that the exponent is always one less than the bounce number.

This pattern continues.

The downward distance for the twelfth bounce is $32 \cdot 0.8^{11}$ feet.

The sum of all the downward distances, then, is:

$$32 + 32 \cdot 0.8 + 32 \cdot 0.8^2 + 32 \cdot 0.8^3 + \cdots + 32 \cdot 0.8^{11}$$

This is a finite geometric series with $a_1 = 32$, $r = 0.8$, and $n = 12$. The formula for the sum of a geometric series is given in the reference sheet provided on the Regents.

$$S_n = \frac{a_1 - a_1 r^n}{1 - r}$$

$$= \frac{32 - 32 \cdot 0.8^{12}}{1 - 0.8}$$

$$\approx 149 \text{ feet}$$

The correct choice is **(4)**.

22. Of the voters who were polled, 264 out of 550 (approximately 48%) said they would vote for Mayor Ortega. Each of the choices needs to be analyzed separately to see which is least appropriate.

Choice (1): The percent of people who vote for Mayor Ortega is not the same as the chance that Mayor Ortega will win. For example, if only 10% of the sample said they were going to vote for Mayor Ortega, there would be closer to a 0% chance that Mayor Ortega would win. Although there is some relationship between the two percentages, they are not the same thing.

Choice (2): The percent based on just the 550 people polled is known as the point estimate, so this is accurate.

Choice (3): The 48% found in the poll is not likely to be the exact percent of people who vote for Mayor Ortega in a full election. However, assuming that the sample size is a reasonable size, the percent is likely be near 48%. There really isn't enough information to say that the actual voter range from 48% is "likely" to be ±4%, but this is still more of an appropriate conclusion than the one from choice (1).

Choice (4): Since so little information is given in the problem—we don't know the total number of voters, for example—it is really hard to make any conclusions based on the poll. It is then appropriate to say that no inference can be made about who will ultimately win the election.

The correct choice is **(1)**.

23. First use the property of exponents that $\left(\dfrac{a}{b}\right)^x = \dfrac{a^x}{b^x}$:

$$\left(\frac{-54x^9}{y^4}\right)^{\frac{2}{3}} = \frac{(-54x^9)^{\frac{2}{3}}}{\left(y^4\right)^{\frac{2}{3}}}$$

Next use the property of exponents that $(ab)^x = a^x b^x$:

$$\frac{(-54x^9)^{\frac{2}{3}}}{(y^4)^{\frac{2}{3}}} = \frac{(-54)^{\frac{2}{3}}(x^9)^{\frac{2}{3}}}{(y^4)^{\frac{2}{3}}}$$

Next use the property of exponents that $(a^x)^y = a^{xy}$:

$$\frac{(-54)^{\frac{2}{3}}(x^9)^{\frac{2}{3}}}{(y^4)^{\frac{2}{3}}} = \frac{(-54)^{\frac{2}{3}}x^{9 \cdot \frac{2}{3}}}{y^{4 \cdot \frac{2}{3}}} = \frac{(-54)^{\frac{2}{3}}x^6}{y^{\frac{8}{3}}}$$

Next use the property of exponents that $a^{\frac{x}{y}} = \left(\sqrt[y]{a}\right)^x$ or $a^{\frac{x}{y}} = \left(\sqrt[y]{a^x}\right)$:

$$\frac{(-54)^{\frac{2}{3}}x^6}{y^{\frac{8}{3}}} = \frac{\left(\sqrt[3]{-54}\right)^2 x^6}{\sqrt[3]{y^8}}$$

Finally, simplify the radicals by factoring the perfect cubes out of the expressions inside the cube roots, and then simplify:

$$\frac{\left(\sqrt[3]{-54}\right)^2 x^6}{\sqrt[3]{y^8}} = \frac{\left(\sqrt[3]{-27 \cdot 2}\right)^2 x^6}{\sqrt[3]{y^6 y^2}}$$

$$= \frac{\left(-3\sqrt[3]{2}\right)^2 x^6}{y^2 \sqrt[3]{y^2}}$$

$$= \frac{(-3)^2 \sqrt[3]{2}^2 x^6}{y^2 \sqrt[3]{y^2}}$$

$$= \frac{(-3)^2 \sqrt[3]{2^2} x^6}{y^2 \sqrt[3]{y^2}}$$

$$= \frac{9x^6 \sqrt[3]{4}}{y^2 \sqrt[3]{y^2}}$$

The correct choice is **(4)**.

24. The money in the account the day their daughter was born is represented by $a_0 = 1000$.

    After one year, since the interest is compounded annually, the Rickerts receive the 1.8% interest and they deposit an additional $750. Calculating 1.8% interest added to the $1000 can be done two ways. First, calculate $1000 + $1000(0.018) = 1018. Alternatively, just multiply $1000(1.018) = 1018. Using this second approach for calculating the amount of money after the interest is earned, the amount of money after one year—including the interest and the new deposit—can be represented as $a_1 = 1000(1.018) + 750 = 1768.

    After two years, the Rickerts receive 1.8% interest on the money that was in the account after one year and the additional $750 deposit. One way to represent this is with the equation $a_2 = 1768(1.018) + 750 \approx 2550. It could also be expressed in terms of a_1 as $a_2 = a_1(1.018) + 750$.

    Likewise, after three years, the amount of money in the account can be expressed as $a_3 = a_2(1.018) + 750$.

    In general when $n > 1$, the recursive expression for calculating a_n is $a_n = a_{n-1}(1.018) + 750$.

    The correct choice is **(3)**.

PART II

25. Use the property of fractional exponents that says $x^{\frac{n}{d}} = \left(\sqrt[d]{x}\right)^n$:

$$(-8)^{\frac{4}{3}} = \left(\sqrt[3]{-8}\right)^4$$
$$= (-2)^4$$
$$= 16$$

26. Use the property of probability that says $P(A \text{ and } B) = P(A) \cdot P(B \text{ given } A)$. The event A is the volunteer receiving the drug. The event B is the patient getting well.

    $P(A \text{ and } B)$ is given as $40\% = 0.4$.

    $P(A)$ is given as $50\% = 0.5$.

    $P(A \text{ and } B) = P(A) \cdot P(B \text{ given } A)$.

$$0.4 = 0.5 \cdot P(B \text{ given } A)$$
$$\frac{0.4}{0.5} = \frac{0.5 \cdot P(B \text{ given } A)}{0.5}$$
$$\frac{0.4}{0.5} = P(B \text{ given } A)$$
$$0.8 = P(B \text{ given } A)$$

27. Simplify each side of the equation, and compare the results:

$$(x^2 + y^2)^2 = (x^2 - y^2)^2 + (2xy)^2$$

For the left-hand side:

$$\left(x^2 + y^2\right)^2 =$$
$$\left(x^2 + y^2\right)\left(x^2 + y^2\right) =$$
$$x^4 + x^2y^2 + x^2y^2 + y^4 =$$
$$x^4 + 2x^2y^2 + y^4 =$$

For the right-hand side:

$$= \left(x^2 - y^2\right)^2 + \left(2xy\right)^2$$
$$= \left(x^2 - y^2\right)\left(x^2 - y^2\right) + \left(2xy\right)^2$$
$$= x^4 - x^2y^2 - x^2y^2 + y^4 + 4x^2y^2$$
$$= x^4 - 2x^2y^2 + y^4 + 4x^2y^2$$
$$= x^4 + 2x^2y^2 + y^4$$

Both sides simplify to the same expression, making the identity true.

28. Since there are six different flavors of jelly beans and there are six sides on the die, you could simulate taking a jelly bean out of the bag by rolling one die. Assign one of the possible numbers, like number 1, to correspond to picking a black jelly bean from the bag.

To see how likely it is to pick four jelly beans at random and have all of them be black, you can roll the die four times and record how many 1s there were. Sometimes there will be no 1s. Sometimes there will be one 1. Sometimes there will be two 1s. Sometimes there will be three 1s. Sometimes there will be four 1s.

Do this experiment 62 times, which will require 248 dice rolls. Find the percent of times out of the 62 experiments that there were four 1s. That will be approximately the probability of getting four black jelly beans from the bag.

29. In this sequence, each term after the first one is one more than double the previous term.

As a recursive formula, this would be:

$$a_1 = 4$$
$$a_n = 2a_{n-1} + 1 \text{ for } n > 1$$

To find the eighth term, the sixth and seventh terms need to be found:

$$
\begin{aligned}
a_6 &= 2a_{6-1} + 1 \\
&= 2a_5 + 1 \\
&= 2 \cdot 79 + 1 \\
&= 159
\end{aligned}
$$

$$
\begin{aligned}
a_7 &= 2a_{7-1} + 1 \\
&= 2a_6 + 1 \\
&= 2 \cdot 159 + 1 \\
&= 319
\end{aligned}
$$

$$
\begin{aligned}
a_8 &= 2a_{8-1} + 1 \\
&= 2a_7 + 1 \\
&= 2 \cdot 319 + 1 \\
&= 639
\end{aligned}
$$

30. The formula for exponential growth is $A_n = A_0 (1 + r)^n$, where A_0 is the starting value, r is the rate of growth, n is the amount of time, and A_n is the ending value.

For this problem, use the following:

$$8.75 = 1.25(1+r)^{49}$$

$$\frac{8.75}{1.25} = \frac{1.25(1+r)^{49}}{1.25}$$

$$7 = (1+r)^{49}$$

$$\sqrt[49]{7} = \sqrt[49]{(1+r)^{49}}$$

$$1.0405 \approx 1 + r$$

$$-1 = -1$$

$$0.040 \approx r$$

The rate of growth is approximately 4%.

31. An odd function is one where $f(-x) = -f(x)$. An even function is one where $f(-x) = f(x)$.

    Calculate $j(-x)$, and compare it to $j(x)$ and $-j(x)$:

    $$j(-x) = (-x)^4 - 3(-x)^2 - 4$$
    $$= x^4 - 3x^2 - 4$$

    $$j(x) = x^4 - 3x^2 - 4$$

    $$-j(x) = -(x^4 - 3x^2 - 4)$$
    $$= -x^4 - 3x^2 + 4$$

    Since $j(-x) = j(x)$, the function is even.

    Since $j(-x) = j(x)$, the function is not odd.

32. If a polynomial function has a factor of $(x - a)$, the graph of that function will have an x-intercept at $(a, 0)$. If it has a factor of $(x - b)$, the graph will have an x-intercept at $(b, 0)$. If the polynomial function has a factor of $(x + c)$, the graph will have an x-intercept at $(-c, 0)$.

    Make a curve that passes through these three points. It will look a little like an "N" and will have a y-intercept $(0, d)$ where d is also positive.

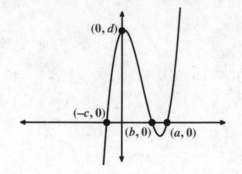

A simpler way to solve this is to pick positive values for a, b, and c with $a > b$, such as $a = 3$, $b = 2$, and $c = 1$. Then use the graphing calculator to make a graph of $p(x) = (x - 3)(x - 2)(x + 1)$.

For the TI-84:

For the TI-Nspire:

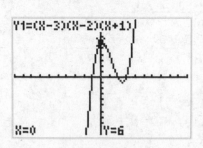

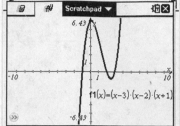

PART III

33. Multiply both sides of the equation by the LCM of all the denominators. In this case, the LCM is $(p-5)(p+3)$:

$$\frac{3p}{p-5} - \frac{2}{p+3} = \frac{p}{p+3}$$

$$\frac{(p-5)(p+3)3p}{p-5} - \frac{(p-5)(p+3)2}{p+3} = \frac{(p-5)(p+3)p}{p+3}$$

$$(p+3)3p - (p-5)2 = (p-5)p$$

$$3p^2 + 9p - 2p + 10 = p^2 - 5p$$

$$3p^2 + 7p + 10 = p^2 - 5p$$

To solve this quadratic equation, move all the terms to one side of the equation and try to factor:

$$3p^2 + 7p + 10 = p^2 - 5p$$

$$\underline{-p^2 + 5p = -p^2 + 5p}$$

$$2p^2 + 12p + 10 = 0$$

$$2\left(p^2 + 6p + 5\right) = 0$$

$$2(p+1)(p+5) = 0$$

$$\begin{array}{ccc} p+1=0 & & p+5=0 \\ \underline{-1=-1} & \text{or} & \underline{-5=-5} \\ p=-1 & & p=-5 \end{array}$$

This equation has two solutions: $p = -1$ and $p = -5$.

34. In this arithmetic sequence, the position in the sequence is the number of days late and the number in that position is the amount of money owed.

It is given that $a_5 = 2.25$ and $a_{21} = 6.25$. To find the value of d, the constant difference from one term to the next, divide the change from term 5 to term 21 by 16 because the common difference must be added 16 times to get from term 5 to term 21:

$$d = \frac{6.25 - 2.25}{21 - 16}$$
$$= \frac{4}{16}$$
$$= 0.25$$

The formula for the nth term of an arithmetic sequence, which can be found in the reference sheet in the back of the Regents, is $a_n = a_1 + (n - 1)d$, where a_n is the nth term, a_1 is the first term, and d is the constant difference.

You can use the given information that $a_5 = 2.25$ together with the value of d that was just calculated to find the value of a_1, which is needed for the formula:

$$a_n = a_1 + (n - 1)d$$
$$a_5 = a_1 + (5 - 1)d$$
$$2.25 = a_1 + 4 \cdot 0.25$$
$$2.25 = a_1 + 1$$
$$\underline{-1 = \quad -1}$$
$$1.25 = a_1$$

So the formula for the nth term is $a_n = 1.25 + (n - 1) \cdot 0.25$.

Use the formula to find the 60th term:

$$a_{60} = 1.25 + (60 - 1) \cdot 0.25$$
$$= 1.25 + 59 \cdot 0.25$$
$$= 1.25 + 14.75$$
$$= 16$$

When the book is 60 days late, the total amount Simon owes will be $16.

35. Start by graphing the horizontal midline $y = -\dfrac{3}{2}$.

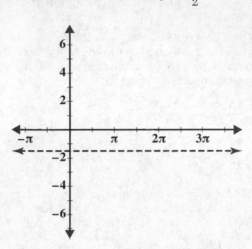

Since the amplitude is given as 2, the highest point will be 2 units above the midline and the lowest point will be 2 units below the midline. Draw horizontal guidelines 2 units above and two units below the midline at

$y = \dfrac{1}{2}$ and $y = -\dfrac{7}{2}$. The curve will fit between these two lines.

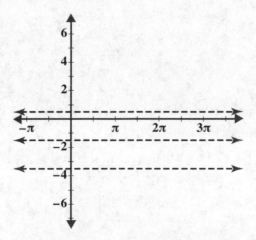

The initial point of one cycle of a sine curve is always on the midline. The simplest place to start is at the point $\left(0, -\frac{3}{2}\right)$. The period is the horizontal distance between the initial point of one cycle of the sine curve and the ending point of that cycle. Since the period is given as 2π, graph the ending point of the cycle at $\left(2\pi, -\frac{3}{2}\right)$.

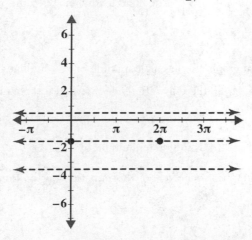

The sine curve passes through the midline halfway between the initial and the ending points. The curve reaches a maximum point at $\frac{1}{4}$ and a minimum point at $\frac{3}{4}$ of the way through the complete cycle. The complete curve looks like this:

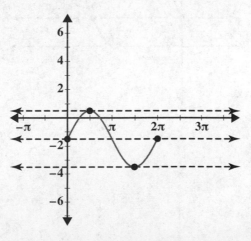

The graph of an equation with the structure $y = A \sin[B(x - C)] + D$ is a sine curve with an amplitude of A, a period of $\dfrac{2\pi}{B}$, a midline of $y = D$, and a horizontal shift to the right of C.

The graph of $y = 2\sin\left(x - \dfrac{\pi}{3}\right) - \dfrac{3}{2}$ has an amplitude of 2, a midline of $y = -\dfrac{3}{2}$, a period of $\dfrac{2\pi}{1} = 2\pi$, and a shift of $\dfrac{\pi}{3}$ to the right compared to $y = 2\sin(x) - \dfrac{3}{2}$.

The original graph also has an amplitude of 2, a midline of $y = -\dfrac{3}{2}$, and a period of 2π. So the only difference between a sketch of the equation $y = 2\sin\left(x - \dfrac{\pi}{3}\right) - \dfrac{3}{2}$ and the sketch from part (a) is that the sketch of the equation in part (b) is shifted to the right by $\dfrac{\pi}{3}$.

Here is a graph of one cycle of this equation on the same graph as the sketch from part (a):

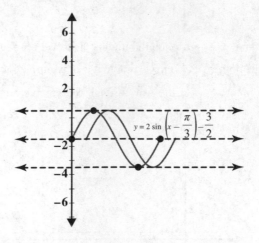

36. On your graphing calculator, enter the data and use the exponential regression feature.

For the TI-84:

Press [STAT] [1] and enter the x-values into L1 and the y-values into L2. Press [STAT] [RIGHT] [0], then select the "Calculate" option.

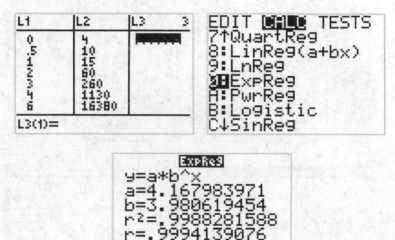

For the TI-Nspire:

From the home screen, select "Lists & Spreadsheet." Enter the data into columns A and B. Press [menu] [4] [A] to select the Exponential Regression function. For X List, enter "x," and for Y List enter "y." Press the "OK" button.

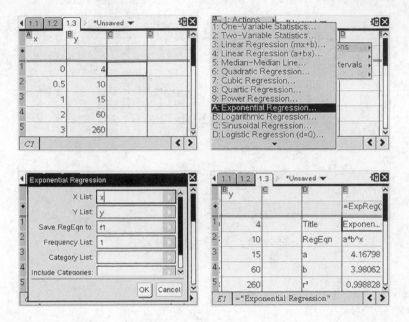

The exponential regression equation is $y = 4.168(3.981)^x$.

Substitute $y = 100$ into the equation, and solve for x using logarithms:

$$100 = 4.168(3.981)^x$$

$$\frac{100}{4.168} = \frac{4.168(3.981)^x}{4.168}$$

$$24 = 3.981^x$$

$$x = \log_{3.981} 24$$

$$\approx 2.3$$

Since the question asks you to round to the nearest quarter hour and 2.3 is closer to 2.25 than it is to 2.5, the answer is 2.25 hours.

PART IV

37. On the x-axis, make every box equal to $\frac{1}{4}$ units so the largest value is 5.

    On the y-axis, make every box equal to 1500 units so the largest value is 30,000.

    To plot the graph, find the values of $V(t)$ and $Z(t)$ for $t = 0, 1, 2, 3, 4,$ and 5.

t	$V(t)$	$Z(t)$
0	28,483	22,151
1	19,482	17,234
2	13,326	13,408
3	9115	10,431
4	6235	8116
5	4264	6314

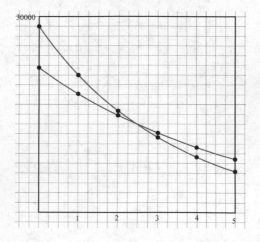

To find the t value for which $V(t) = Z(t)$, use the graphing calculator to find the x-coordinate of the intersection point of the two curves.

For the TI-84:   For the TI-Nspire:

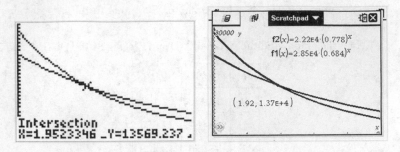

Rounded to the nearest hundredth, $V(t) = Z(t)$ when $t \approx 1.95$ (or on the TI-Nspire because of rounding issues, 1.92). This means that after approximately 1.95 years (or with the TI-Nspire, 1.92 years) the value of the car is the same as the amount Zach still owes on the loan.

Find the value of t for which $V(t) = 3000$:

$$3000 = 28482.698(0.684)^t$$

$$\frac{3000}{28482.698} = \frac{28482.698(0.684)^t}{28482.698}$$

$$0.1053271 = 0.684^t$$

$$t = \log_{0.684} 0.1053271$$

$$= 5.92$$

$$\approx 6$$

It will take approximately 6 years for Zach to cancel this policy.

Topic	Question Numbers	Number of Points	Your Points	Your Percentage
1. Polynomial Expressions and Equations	6, 8, 12, 13, 15, 20, 27, 32	2 + 2 + 2 + 2 + 2 + 2 + 2 + 2 = 16		
2. Complex Numbers	2, 3	2 + 2 = 4		
3. Exponential Expressions and Equations	9, 10, 30, 36, 37	2 + 2 + 2 + 4 + 6 = 16		
4. Rational Expressions and Equations	33	4		
5. Radical Expresions and Equations	4, 23, 25	2 + 2 + 2 = 6		
6. Trigonometric Expressions and Equations	5, 7, 18, 35	2 + 2 + 2 + 4 = 10		
7. Functions	1, 14, 31	2 + 2 + 2 = 6		
8. Systems of Equations	19	2		
9. Sequences and Series	21, 24, 29, 34	2 + 2 + 2 + 4 = 10		
10. Probability	26	2		
11. Statistics	11, 16, 17, 22, 28	2 + 2 + 2 + 2 + 2 = 10		

HOW TO CONVERT YOUR RAW SCORE TO YOUR ALGEBRA II REGENTS EXAMINATION SCORE

The accompanying conversion chart must be used to determine your final score on the August 2017 Regents Examination in Algebra II. To find your final exam score, locate in the column labeled "Raw Score" the total number of points you scored out of a possible 86 points. Since partial credit is allowed in Parts II, III, and IV of the test, you may need to approximate the credit you would receive for a solution that is not completely correct. Then locate in the adjacent column to the right the scale score that corresponds to your raw score. The scale score is your final Algebra II Regents Examination score.

Regents Examination in Algebra II—August 2017
Chart for Converting Total Test Raw Scores to Final
Examination Scores (Scaled Scores)

Raw Score	Scale Score	Performance Level	Raw Score	Scale Score	Performance Level	Raw Score	Scale Score	Performance Level
86	100	5	57	82	4	28	67	3
85	99	5	56	81	4	27	66	3
84	98	5	55	81	4	26	66	3
83	97	5	54	80	4	25	65	3
82	96	5	53	80	4	24	63	2
81	96	5	52	80	4	23	62	2
80	95	5	51	79	4	22	61	2
79	94	5	50	79	4	21	60	2
78	93	5	49	79	4	20	58	2
77	93	5	48	78	4	19	55	2
76	92	5	47	78	4	18	54	1
75	91	5	46	78	4	17	53	1
74	91	5	45	77	3	16	51	1
73	90	5	44	77	3	15	49	1
72	89	5	43	76	3	14	47	1
71	89	5	42	76	3	13	45	1
70	88	5	41	76	3	12	43	1
69	87	5	40	75	3	11	40	1
68	87	5	39	75	3	10	37	1
67	86	5	38	74	3	9	34	1
66	86	5	37	74	3	8	31	1
65	86	5	36	73	3	7	28	1
64	85	5	35	73	3	6	25	1
63	84	4	34	72	3	5	21	1
62	84	4	33	71	3	4	17	1
61	83	4	32	71	3	3	13	1
60	83	4	31	70	3	2	9	1
59	82	4	30	69	3	1	5	1
58	82	4	29	68	3	0	0	1

Examination
June 2018
Algebra II

HIGH SCHOOL MATH REFERENCE SHEET

Conversions

1 inch = 2.54 centimeters

1 meter = 39.37 inches

1 mile = 5280 feet

1 mile = 1760 yards

1 mile = 1.609 kilometers

1 kilometer = 0.62 mile

1 pound = 16 ounces

1 pound = 0.454 kilogram

1 kilogram = 2.2 pounds

1 ton = 2000 pounds

1 cup = 8 fluid ounces

1 pint = 2 cups

1 quart = 2 pints

1 gallon = 4 quarts

1 gallon = 3.785 liters

1 liter = 0.264 gallon

1 liter = 1000 cubic centimeters

Formulas

Triangle	$A = \frac{1}{2}bh$
Parallelogram	$A = bh$
Circle	$A = \pi r^2$
Circle	$C = \pi d$ or $C = 2\pi r$

Formulas (continued)

General Prisms	$V = Bh$
Cylinder	$V = \pi r^2 h$
Sphere	$V = \frac{4}{3}\pi r^3$
Cone	$V = \frac{1}{3}\pi r^2 h$
Pyramid	$V = \frac{1}{3}Bh$
Pythagorean Theorem	$a^2 + b^2 = c^2$
Quadratic Formula	$x = \dfrac{-b \pm \sqrt{b^2 - 4ac}}{2a}$
Arithmetic Sequence	$a_n = a_1 + (n - 1)d$
Geometric Sequence	$a_n = a_1 r^{n-1}$
Geometric Series	$S_n = \dfrac{a_1 - a_1 r^n}{1 - r}$ where $r \neq 1$
Radians	$1 \text{ radian} = \dfrac{180}{\pi} \text{ degrees}$
Degrees	$1 \text{ degree} = \dfrac{\pi}{180} \text{ radians}$
Exponential Growth/Decay	$A = A_0 e^{k(t - t_0)} + B_0$

PART I

Answer all 24 questions in this part. Each correct answer will receive 2 credits. No partial credit will be allowed. For each statement or question, write in the space provided the numeral preceding the word or expression that best completes the statement or answers the question.　[48 credits]

1 The graphs of the equations $y = x^2 + 4x - 1$ and $y + 3 = x$ are drawn on the same set of axes. One solution of this system is

(1) $(-5, -2)$ (3) $(1, 4)$

(2) $(-1, -4)$ (4) $(-2, -1)$ 1 _____

2 Which statement is true about the graph of $f(x) = \left(\dfrac{1}{8}\right)^x$?

(1) The graph is always increasing.

(2) The graph is always decreasing.

(3) The graph passes through $(1, 0)$.

(4) The graph has an asymptote, $x = 0$.   2 _____

3 For all values of x for which the expression is defined, $\dfrac{x^3 + 2x^2 - 9x - 18}{x^3 - x^2 - 6x}$, in simplest form, is equivalent to

(1) 3 (3) $\dfrac{x+3}{x}$

(2) $-\dfrac{17}{2}$ (4) $\dfrac{x^2-9}{x(x-3)}$ 3 _____

4 A scatterplot showing the weight, w, in grams, of each crystal after growing t hours is shown below.

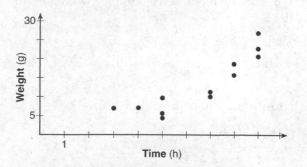

The relationship between weight, w, and time, t, is best modeled by

(1) $w = 4^t + 5$　　　　　　　(3) $w = 5(2.1)^t$

(2) $w = (1.4)^t + 2$　　　　　(4) $w = 8(.75)^t$　　　4 _____

5 Where i is the imaginary unit, the expression $(x + 3i)^2 - (2x - 3i)^2$ is equivalent to

(1) $-3x^2$　　　　　　　　　(3) $-3x^2 + 18xi$

(2) $-3x^2 - 18$　　　　　　　(4) $-3x^2 - 6xi - 18$　　　5 _____

6 Which function is even?

(1) $f(x) = \sin x$　　　　　　(3) $f(x) = |x - 2| + 5$

(2) $f(x) = x^2 - 4$　　　　　　(4) $f(x) = x^4 + 3x^3 + 4$　　　6 _____

7 The function $N(t) = 100e^{-0.023t}$ models the number of grams in a sample of cesium-137 that remain after t years. On which interval is the sample's average rate of decay the fastest?

(1) $[1, 10]$ (3) $[15, 25]$

(2) $[10, 20]$ (4) $[1, 30]$ 7 _____

8 Which expression can be rewritten as $(x + 7)(x - 1)$?

(1) $(x + 3)^2 - 16$

(2) $(x + 3)^2 - 10(x + 3) - 2(x + 3) + 20$

(3) $\dfrac{(x-1)\left(x^2 - 6x - 7\right)}{(x+1)}$

(4) $\dfrac{(x+7)\left(x^2 + 4x + 3\right)}{(x+3)}$ 8 _____

9 What is the solution set of the equation $\dfrac{2}{x} - \dfrac{3x}{x+3} = \dfrac{x}{x+3}$?

(1) $\{3\}$ (3) $\{-2, 3\}$

(2) $\left\{\dfrac{3}{2}\right\}$ (4) $\left\{-1, \dfrac{3}{2}\right\}$ 9 _____

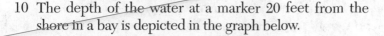

10 The depth of the water at a marker 20 feet from the shore in a bay is depicted in the graph below.

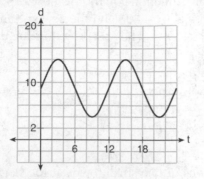

If the depth, d, is measured in feet and time, t, is measured in hours since midnight, what is an equation for the depth of the water at the marker?

(1) $d = 5 \cos\left(\dfrac{\pi}{6}t\right) + 9$ (3) $d = 9 \sin\left(\dfrac{\pi}{6}t\right) + 5$

(2) $d = 9 \cos\left(\dfrac{\pi}{6}t\right) + 5$ (4) $d = 5 \sin\left(\dfrac{\pi}{6}t\right) + 9$ 10 _____

11 On a given school day, the probability that Nick oversleeps is 48% and the probability he has a pop quiz is 25%. Assuming these two events are independent, what is the probability that Nick oversleeps and has a pop quiz on the same day?

(1) 73%   (3) 23%

(2) 36%   (4) 12%   11 _____

12 If $x - 1$ is a factor of $x^3 - kx^2 + 2x$, what is the value of k?

(1) 0   (3) 3

(2) 2   (4) −3   12 _____

13 The profit function, $p(x)$, for a company is the cost function, $c(x)$, subtracted from the revenue function, $r(x)$. The profit function for the Acme Corporation is $p(x) = -0.5x^2 + 250x - 300$ and the revenue function is $r(x) = -0.3x^2 + 150x$. The cost function for the Acme Corporation is

(1) $c(x) = 0.2x^2 - 100x + 300$

(2) $c(x) = 0.2x^2 + 100x + 300$

(3) $c(x) = -0.2x^2 + 100x - 300$

(4) $c(x) = -0.8x^2 + 400x - 300$

13 _____

14 The populations of two small towns at the beginning of 2018 and their annual population growth rate are shown in the table below.

Town	Population	Annual Population Growth Rate
Jonesville	1240	6% increase
Williamstown	890	11% increase

Assuming the trend continues, approximately how many years after the beginning of 2018 will it take for the populations to be equal?

(1) 7   (3) 68

(2) 20   (4) 125

14 _____

15 What is the inverse of $f(x) = x^3 - 2$?

(1) $f^{-1}(x) = \sqrt[3]{x} + 2$ (3) $f^{-1}(x) = \sqrt[3]{x+2}$

(2) $f^{-1}(x) = \pm\sqrt[3]{x} + 2$ (4) $f^{-1}(x) = \pm\sqrt[3]{x+2}$

15 _____

16 A 4th degree polynomial has zeros −5, 3, i, and −i.
Which graph could represent the function defined by
this polynomial?

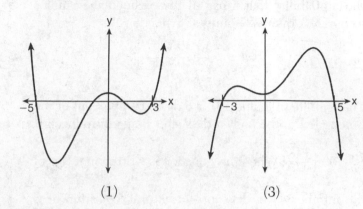

(1)   (3)

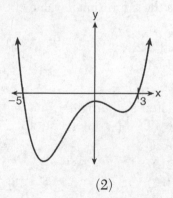

(2)   (4)   16 _____

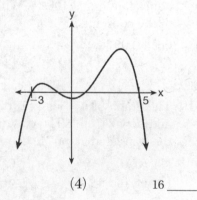

17 The weights of bags of Graseck's Chocolate Candies are normally distributed with a mean of 4.3 ounces and a standard deviation of 0.05 ounces. What is the probability that a bag of these chocolate candies weighs less than 4.27 ounces?

(1) 0.2257   (3) 0.7257

(2) 0.2743   (4) 0.7757   17 _____

18 The half-life of iodine-131 is 8 days. The percent of the isotope left in the body d days after being introduced is $I = 100 \left(\dfrac{1}{2}\right)^{\frac{d}{8}}$. When this equation is written in terms of the number e, the base of the natural logarithm, it is equivalent to $I = 100e^{kd}$. What is the approximate value of the constant, k?

(1) −0.087   (3) −11.542

(2) 0.087   (4) 11.542   18 _____

19 The graph of $y = \log_2 x$ is translated to the right 1 unit and down 1 unit. The coordinates of the x-intercept of the translated graph are

(1) $(0, 0)$ (3) $(2, 0)$

(2) $(1, 0)$ (4) $(3, 0)$ 19 _____

20 For positive values of x, which expression is equivalent to $\sqrt{16x^2} \bullet x^{\frac{2}{3}} + \sqrt[3]{8x^5}$?

(1) $6\sqrt[5]{x^3}$ (3) $4\sqrt[3]{x^2} + 2\sqrt[3]{x^5}$

(2) $6\sqrt[3]{x^5}$ (4) $4\sqrt{x^3} + 2\sqrt[5]{x^3}$ 20 _____

21 Which equation represents a parabola with a focus of $(-2, 5)$ and a directrix of $y = 9$?

   (1) $(y - 7)^2 = 8(x + 2)$
   (2) $(y - 7)^2 = -8(x + 2)$
   (3) $(x + 2)^2 = 8(y - 7)$
   (4) $(x + 2)^2 = -8(y - 7)$ 21 _____

22 Given the following polynomials

$$x = (a + b + c)^2$$
$$y = a^2 + b^2 + c^2$$
$$z = ab + bc + ac$$

Which identity is true?

   (1) $x = y - z$ (3) $x = y - 2z$
   (2) $x = y + z$ (4) $x = y + 2z$ 22 _____

23 On average, college seniors graduating in 2012 could compute their growing student loan debt using the function $D(t) = 29{,}400(1.068)^t$, where t is time in years. Which expression is equivalent to $29{,}400(1.068)^t$ and could be used by students to identify an approximate daily interest rate on their loans?

   (1) $29{,}400\left(1.068^{\frac{1}{365}}\right)^t$ (3) $29{,}400\left(1 + \frac{0.068}{365}\right)^t$

   (2) $29{,}400\left(\frac{1.068}{365}\right)^{365t}$ (4) $29{,}400\left(1.068^{\frac{1}{365}}\right)^{365t}$ 23 _____

24 A manufacturing plant produces two different-sized containers of peanuts. One container weighs x ounces and the other weighs y pounds. If a gift set can hold one of each size container, which expression represents the number of gift sets needed to hold 124 ounces?

(1) $\dfrac{124}{16x+y}$ (3) $\dfrac{124}{x+16y}$

(2) $\dfrac{x+16y}{124}$ (4) $\dfrac{16x+y}{124}$ 24 _____

PART II

Answer all 8 questions in this part. Each correct answer will receive 2 credits. Clearly indicate the necessary steps, including appropriate formula substitutions, diagrams, graphs, charts, etc. For all questions in this part, a correct numerical answer with no work shown will receive only 1 credit.   [16 credits]

25 A survey about television-viewing preferences was given to randomly selected freshmen and seniors at Fairport High School. The results are shown in the table below.

Favorite Type of Program

	Sports	Reality Show	Comedy Series
Senior	83	110	67
Freshman	119	103	54

A student response is selected at random from the results. State the *exact* probability the student response is from a freshman, given the student prefers to watch reality shows on television.

26 On the grid below, graph the function
$f(x) = x^3 - 6x^2 + 9x + 6$ on the domain $-1 \le x \le 4$.

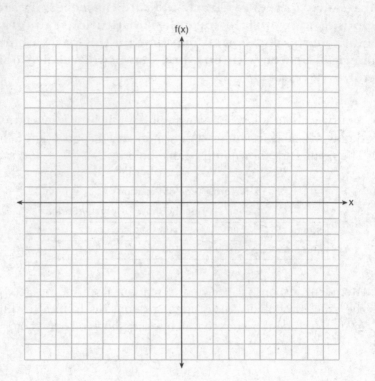

27 Solve the equation $2x^2 + 5x + 8 = 0$. Express the answer in $a + bi$ form.

28 Chuck's Trucking Company has decided to initiate an Employee of the Month program. To determine the recipient, they put the following sign on the back of each truck.

How's My Driving?

Call 1-555-DRIVING

The driver who receives the highest number of positive comments will win the recognition. Explain *one* statistical bias in this data collection method.

29 Determine the quotient and remainder when
$(6a^3 + 11a^2 - 4a - 9)$ is divided by $(3a - 2)$.

Express your answer in the form $q(a) + \dfrac{r(a)}{d(a)}$.

30 The recursive formula to describe a sequence is shown below.

$$a_1 = 3$$
$$a_n = 1 + 2a_{n-1}$$

State the first four terms of this sequence.

Can this sequence be represented using an explicit geometric formula? Justify your answer.

31 The Wells family is looking to purchase a home in a suburb of Rochester with a 30-year mortgage that has an annual interest rate of 3.6%. The house the family wants to purchase is $152,500 and they will make a $15,250 down payment and borrow the remainder. Use the formula below to determine their monthly payment, to the *nearest dollar*.

$$M = \frac{P\left(\dfrac{r}{12}\right)\left(1 + \dfrac{r}{12}\right)^{n}}{\left(1 + \dfrac{r}{12}\right)^{n} - 1}$$

M = monthly payment
P = amount borrowed
r = annual interest rate
n = total number of monthly payments

32 An angle, θ, is in standard position and its terminal side passes through the point $(2, -1)$. Find the *exact* value of sin θ.

PART III

Answer all 4 questions in this part. Each correct answer will receive 4 credits. Clearly indicate the necessary steps, including appropriate formula substitutions, diagrams, graphs, charts, etc. For all questions in this part, a correct numerical answer with no work shown will receive only 1 credit.   [16 credits]

33 Solve algebraically for all values of x:

$$\sqrt{6 - 2x} + x = 2(x + 15) - 9$$

34 Joseph was curious to determine if scent improves memory. A test was created where better memory is indicated by higher test scores. A controlled experiment was performed where one group was given the test on scented paper and the other group was given the test on unscented paper. The summary statistics from the experiment are given below.

	Scented Paper	Unscented Paper
$\overline{x}$	23	18
s_x	2.898	2.408

Calculate the difference in means in the experimental test grades (scented – unscented).

A simulation was conducted in which the subjects' scores were rerandomized into two groups 1000 times. The differences of the group means were calculated each time. The results are shown below.

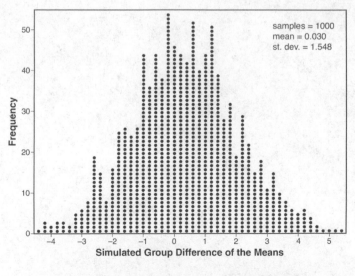

samples = 1000
mean = 0.030
st. dev. = 1.548

Simulated Group Difference of the Means

Question 34 is continued on the next page.

Question 34 continued.

Use the simulation results to determine the interval representing the middle 95% of the difference in means, to the *nearest hundredth*.

Is the difference in means in Joseph's experiment statistically significant based on the simulation? Explain.

35 Carla wants to start a college fund for her daughter Lila. She puts \$63,000 into an account that grows at a rate of 2.55% per year, compounded monthly. Write a function, $C(t)$, that represents the amount of money in the account t years after the account is opened, given that no more money is deposited into or withdrawn from the account.

Calculate algebraically the number of years it will take for the account to reach \$100,000, to the *nearest hundredth of a year*.

36 The height, $h(t)$ in cm, of a piston, is given by the equation $h(t) = 12 \cos\left(\dfrac{\pi}{3}t\right) + 8$, where t represents the number of seconds since the measurements began.

Determine the average rate of change, in cm/sec, of the piston's height on the interval $1 \leq t \leq 2$.

At what value(s) of t, to the *nearest tenth of a second*, does $h(t) = 0$ in the interval $1 \leq t \leq 5$? Justify your answer.

PART IV

Answer the question in this part. A correct answer will receive 6 credits. Clearly indicate the necessary steps, including appropriate formula substitutions, diagrams, graphs, charts, etc. A correct numerical answer with no work shown will receive only 1 credit. [6 credits]

37 Website popularity ratings are often determined using models that incorporate the number of visits per week a website receives. One model for ranking websites is $P(x) = \log(x - 4)$, where x is the number of visits per week in thousands and $P(x)$ is the website's popularity rating.

According to this model, if a website is visited 16,000 times in one week, what is its popularity rating, rounded to the *nearest tenth*?

Graph $y = P(x)$ on the axes below.

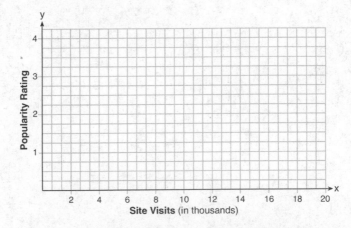

Question 37 is continued on the next page.

Question 37 continued.

An alternative rating model is represented by $R(x) = \frac{1}{2}x - 6$, where x is the number of visits per week in thousands. Graph $R(x)$ on the same set of axes. For what number of weekly visits will the two models provide the same rating?

Answers
June 2018
Algebra II

Answer Key

PART I

1. (2)	**5.** (3)	**9.** (4)	**13.** (1)	**17.** (2)	**21.** (4)
2. (2)	**6.** (2)	**10.** (4)	**14.** (1)	**18.** (1)	**22.** (4)
3. (3)	**7.** (1)	**11.** (4)	**15.** (3)	**19.** (4)	**23.** (4)
4. (2)	**8.** (1)	**12.** (3)	**16.** (2)	**20.** (2)	**24.** (3)

PART II

25. $\dfrac{103}{213}$

26.

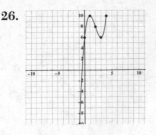

27. $-\dfrac{5}{4} \pm \dfrac{\sqrt{39}}{4} i$

28. Not a random sampling

29. $2a^2 + 5a + 2 - \dfrac{5}{3a-2}$

30. 3, 7, 15, 31; this is not a geometric series.

31. $624

32. $\dfrac{-1}{\sqrt{5}}$

PART III

33. -15

34. 5; between -3.07 and 3.13. It is significant because $5 > 3.13$.

35. $C(t) = 63{,}000\left(1 + \dfrac{0.0255}{12}\right)^{12t}$; 18.14

36. -12; zeros at 2.2 and 3.8

PART IV

37. 1.1; 14,000 visits

In **Parts II–IV**, you are required to show how you arrived at your answers. For sample methods of solutions, see the *Answers Explained* section.

Answers Explained

PART I

1. Rewrite the second equation as $y = x - 3$. Using the substitution method, replace the y in the first equation with the expression $x - 3$ and solve:

$$x - 3 = x^2 + 4x - 1$$
$$\underline{-x \quad\quad = \quad\quad -x}$$
$$-3 = x^2 + 3x - 1$$
$$\underline{+3 = \quad\quad\quad +3}$$
$$0 = x^2 + 3x + 2$$
$$0 = (x + 2)(x + 1)$$

$$
\begin{array}{ccc}
x + 2 = 0 & & x + 1 = 0 \\
\underline{-2 = -2} & \text{or} & \underline{-1 = -1} \\
x = -2 & & x = -1
\end{array}
$$

For the $x = -2$ solution, the y-value can be calculated by plugging $x = -2$ into either original equation:

$$y + 3 = -2$$
$$\underline{-3 = -3}$$
$$y = -5$$

For the $x = -1$ solution, the y-value can be calculated by plugging $x = -1$ into either original equation:

$$y + 3 = -1$$
$$\underline{-3 \quad -3}$$
$$y = -4$$

The two solutions are $(-2, -5)$ and $(-1, -4)$.

This question can also be solved by graphing the two equations on the graphing calculator and then using the intersect feature to find the intersection points.

For the TI-84:   For the TI-Nspire:

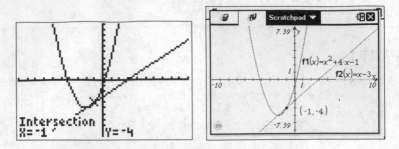

The correct choice is (**2**).

2. An exponential equation of the form $y = b^x$ will model exponential growth and will always be increasing if b is greater than 1. An exponential function of the form $y = b^x$ will model exponential decay and will always be decreasing if b is between 0 and 1.

This can also be seen by graphing the equation on the graphing calculator.

For the TI-84:   For the TI-Nspire:

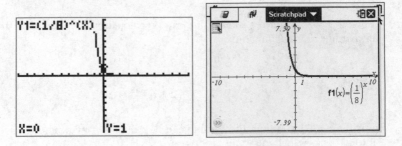

The graph shows that the function is always decreasing.

The correct choice is (**2**).

3. Factor the numerator and the denominator. Then cancel like terms. The numerator can be factored by grouping. The terms in the denominator have a common factor of x:

$$\frac{x^3 + 2x^2 - 9x - 18}{x^3 - x^2 - 6x} =$$

$$\frac{x^2(x+2) - 9(x+2)}{x(x^2 - x - 6)} =$$

$$\frac{(x+2)(x^2 - 9)}{x(x-3)(x+2)} =$$

$$\frac{(x+2)(x-3)(x+3)}{x(x-3)(x+2)} = \frac{x+3}{x}$$

The correct choice is **(3)**.

4. The scatterplot fits a curve with exponential growth. Equations of exponential growth have the form $y = a \cdot b^x + c$, where $b > 1$. Since choice (4) has a b-value of .75, it can be eliminated.

By plugging $t = 10$ into the other three choices, you can see which one results in a w-value of 30.

Choice (1):

$$4^{10} + 5 = 1{,}048{,}581$$

Choice (2):

$$(1.4)^{10} + 2 \approx 31$$

Choice (3):

$$5(2.1)^{10} \approx 8340$$

If you set the window of the graphing calculator to match the window of the diagram, you can see that the graph of choice (2) closely matches the diagram.

For the TI-84:

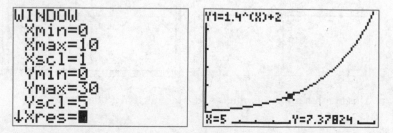

For the TI-Nspire:

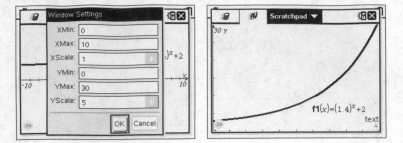

The correct choice is (**2**).

5. The imaginary unit, i, has the property that $i^2 = -1$. Simplify the given expression:

$$(x+3i)^2 - (2x-3i)^2 =$$
$$x^2 + 3ix + 3xi + 9i^2 - (4x^2 - 6ix - 6ix + 9i^2) =$$
$$x^2 + 6ix - 9 - (4x^2 - 12ix - 9) =$$
$$x^2 + 6ix - 9 - 4x^2 + 12ix + 9 = -3x^2 + 18xi$$

The correct choice is (**3**).

6. The graph of an even function is symmetric about the y-axis.

Choice (1):

For the TI-84:   For the TI-Nspire:

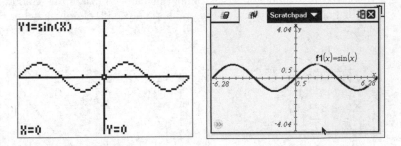

This graph has origin symmetry, so the function is odd, not even.

Choice (2):

For the TI-84:

For the TI-Nspire:

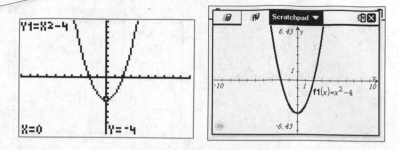

This graph is symmetric about the y-axis, so the function is even.

Choice (3):

For the TI-84:

For the TI-Nspire:

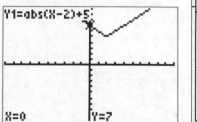

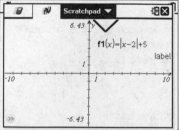

This graph has symmetry about the line $x = 2$ but not about the y-axis, so the function is neither even nor odd.

Choice (4):

For the TI-84:   For the TI-Nspire:

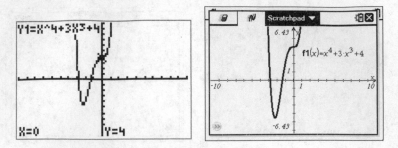

This graph does not have any symmetry, so the function is neither even nor odd.

The correct choice is **(2)**.

7. The average rate of change of the function N on the interval $[a, b]$ can be calculated with the formula $\dfrac{N(b) - N(a)}{b - a}$.

Choice (1):
$$\frac{N(10) - N(1)}{10 - 1} \approx \frac{79.453 - 97.726}{10 - 1} \approx -2.03$$

Choice (2):
$$\frac{N(20) - N(10)}{20 - 10} \approx \frac{63.128 - 79.453}{20 - 10} \approx -1.63$$

Choice (3):
$$\frac{N(25) - N(15)}{25 - 15} \approx \frac{56.27 - 70.822}{25 - 15} \approx -1.46$$

Choice (4):
$$\frac{N(30) - N(1)}{30 - 1} \approx \frac{50.158 - 97.726}{30 - 1} \approx -1.64$$

The correct choice is **(1)**.

8. First simplify the given expression.

$$(x+7)(x-1)$$
$$= x^2 - 1x + 7x - 7$$
$$= x^2 + 6x - 7$$

Then check the answer choices until you find one that simplifies to the same expression.

Choice (1):

$$(x+3)^2 - 16$$
$$= x^2 + 6x + 9 - 16$$
$$= x^2 + 6x - 7$$

Choice (2):

$$(x+3)^2 - 10(x+3) - 2(x+3) + 20$$
$$= x^2 + 6x + 9 - 10x - 30 - 2x - 6 + 20$$
$$= x^2 - 6x - 7$$

Choice (3):

$$\frac{(x-1)(x^2 - 6x - 7)}{(x+1)}$$
$$= \frac{(x-1)(x-7)(x+1)}{(x+1)}$$
$$= (x-1)(x-7)$$
$$= x^2 - 8x + 7$$

Choice (4):

$$\frac{(x+7)(x^2 + 4x + 3)}{(x+3)}$$
$$= \frac{(x+7)(x+3)(x+1)}{(x+3)}$$
$$= (x+7)(x+1)$$
$$= x^2 + 8x + 7$$

The correct choice is (**1**).

9. Multiply both sides of the equation by the least common multiple of all the denominators, $x(x + 3)$. Then solve the resulting quadratic equation:

$$\frac{2}{x} - \frac{3x}{x+3} = \frac{x}{x+3}$$

$$x(x+3)\left(\frac{2}{x} - \frac{3x}{x+3}\right) = x(x+3)\frac{x}{x+3}$$

$$x(x+3)\frac{2}{x} - x(x+3)\frac{3x}{x+3} = x(x+3)\frac{x}{x+3}$$

$$(x+3)2 - x \cdot 3x = x \cdot x$$

$$2x + 6 - 3x^2 = x^2$$

$$\underline{-2x - 6 + 3x^2 = -2x - 6 + 3x^2}$$

$$0 = 4x^2 - 2x - 6$$

$$0 = 2(2x^2 - x - 3)$$

$$0 = 2(2x - 3)(x + 1)$$

$$
\begin{array}{ccc}
2x - 3 = 0 & & x + 1 = 0 \\
\underline{+3 = +3} & \text{or} & \underline{-1 = -1} \\
2x = 3 & & x = -1 \\
x = \dfrac{3}{2} & &
\end{array}
$$

Check to see if either of these answers needs to be eliminated. Since neither of these values would make any of the original denominators equal to zero, no answer needs to be eliminated.

The correct choice is **(4)**.

10. The equation of a graph like this is going to be either in the form $y = A\cos(Bx) + D$ or in the form $y = A\sin(Bx) + D$, where A is the amplitude, B is the frequency, and D is the vertical shift.

To calculate A, find the difference between the maximum point and the minimum point. Divide that difference by 2. For this graph, the maximum point is at 14 and the minimum point is at 4. The difference between these is 10, so the value of A is 5.

If the y-intercept of the curve is at a maximum or a minimum, it is a cosine curve. If the y-intercept is at a point in the middle of the curve, it is a sine curve. Since the y-intercept is in the middle of the curve shown, this must be a sine equation.

To calculate B, divide 2π by the distance from one maximum to the next maximum. One maximum is at $(3, 14)$, and the next one is at $(15, 14)$. The distance between these is 12, so the value of B is $\dfrac{2\pi}{12} = \dfrac{\pi}{6}$.

To calculate the value of D, find the average of the maximum point and the minimum point of the curve. The maximum is 14 and the minimum is 4, so $D = \dfrac{14+4}{2} = \dfrac{18}{2} = 9$.

The equation, then, is $d = 5\sin\left(\dfrac{\pi}{6}t\right) + 9$.

The correct choice is **(4)**.

11. When two events are independent, the probability of both things happening is the product of the probabilities of each of them happening. Since these events are independent, the probability of them both happening is:

$$0.48 \times 0.25 = .12 = 12\%$$

The correct choice is **(4)**.

12. The factor theorem says that if $x - a$ is a factor of polynomial $f(x)$, then $f(a) = 0$. Since $a = 1$ in this example, this means that $1^3 - k \cdot 1^2 + 2 \cdot 1 = 0$. Solve this equation for k:

$$1^3 - k \cdot 1^2 + 2 \cdot 1 = 0$$
$$1 - k + 2 = 0$$
$$3 - k = 0$$
$$\underline{+k = +k}$$
$$3 = k$$

Another way to answer this question is to factor out an x from the polynomial $x(x^2 - kx + 2)$. Since it is given that $x - 1$ is a factor, the second factor of this expression must factor to $(x - 1)(x - 2)$ in order for there to be a $+ 2$ constant in the product.

Since $(x - 1)(x - 2)$ multiplies to $x^2 - 3x + 2$, then k must be 3.

The correct choice is **(3)**.

13. From the first sentence, the equation relating these three functions is
$p(x) = r(x) - c(x)$.

Replace this equation with the given information:

$$-0.5x^2 + 250x - 300 = -0.3x^2 + 150x - c(x)$$

Isolate the $c(x)$ with algebra:

$$-0.5x^2 + 250x - 300 = -0.3x^2 + 150x - c(x)$$
$$\underline{+0.3x^2 \qquad\qquad\qquad = +0.3x^2}$$
$$-0.2x^2 + 250x - 300 = 150x - c(x)$$
$$\underline{-150x \qquad\qquad = -150x}$$
$$-0.2x^2 + 100x - 300 = -c(x)$$
$$\frac{-0.2x^2 + 100x - 300}{-1} = \frac{-c(x)}{-1}$$
$$0.2x^2 - 100x + 300 = c(x)$$

The correct choice is (**1**).

14. Use the exponential growth equation $y = a \cdot (1 + r)^x$, where a is the initial value and r is the growth rate. The equation for Jonesville's population is $y = 1240 \cdot 1.06^x$. The equation for Williamstown's population is $y = 890 \cdot 1.11^x$.

The simplest way to complete this question is to test the four choices to see which value of x makes these two expressions nearly equal.

Choice (1):

$$1240 \cdot 1.06^7 \approx 1865$$
$$890 \cdot 1.11^7 \approx 1848$$

Choice (2):

$$1240 \cdot 1.06^{20} \approx 3977$$
$$890 \cdot 1.11^{20} \approx 7175$$

Choice (3):

$$1240 \cdot 1.06^{68} \approx 65,196$$
$$890 \cdot 1.11^{68} \approx 1,074,862$$

Choice (4):

$$1240 \cdot 1.06^{125} \approx 1,805,738$$
$$890 \cdot 1.11^{125} \approx 411,872,082$$

Another way to solve this is to graph the two equations on a graphing calculator and find the x-coordinate of the intersection point.

For the TI-84:

For the TI-Nspire:

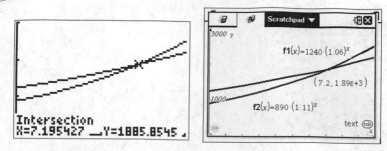

Intersection
X=7.195427 _Y=1885.8545

The correct choice is (1).

15. To calculate the inverse of a function, change the $f(x)$ into an x and the x into a $f^{-1}(x)$. Then use algebra to isolate $f^{-1}(x)$:

$$x = \left[f^{-1}(x) \right]^3 - 2$$
$$\underline{+2 = \qquad\qquad +2}$$
$$x + 2 = \left[f^{-1}(x) \right]^3$$
$$\sqrt[3]{x+2} = \sqrt[3]{\left[f^{-1}(x) \right]^3}$$
$$\sqrt[3]{x+2} = f^{-1}(x)$$

The correct choice is (3).

16. The real-number zeros of a polynomial are also the x-intercepts of the graph of that polynomial. In this question, the polynomial has two real zeros: −5 and 3. Of the four answer choices, only choice (2) has x-intercepts only at (−5, 0) and (3, 0). The two imaginary roots of i and $−i$ do not correspond to any x-intercepts on the graph.

The correct choice is (2).

17. Use the normal cumulative distribution (normal cdf) function on the graphing calculator.

For the TI-84:
Press [2ND] [VARS] [2].

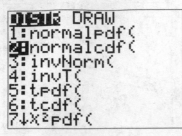

For lower, input 0. For upper, input 4.27. For μ, input the median 4.3. For σ, input 0.05.

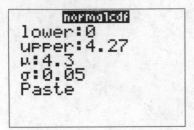

Paste and press [ENTER].

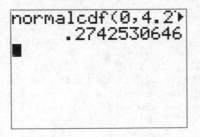

For the TI-NSpire:

From the home screen, press [A] [menu] [6] [5] [2].

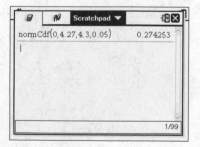

For Lower Bound, input 0. For Upper Bound, input 4.27. For μ, input the median 4.3. For σ, input 0.05.

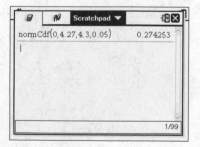

Select and press the OK button.

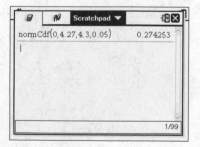

The probability is approximately 0.2743.

The correct choice is **(2)**.

18. Since the two equations are equivalent, $100\left(\dfrac{1}{2}\right)^{\frac{d}{8}} = 100e^{kd}$. The steps for solving for k are as follows.

Step 1: Divide both sides by 100.

$$\frac{100\left(\dfrac{1}{2}\right)^{\frac{d}{8}}}{100} = \frac{100e^{kd}}{100}$$

$$\left(\frac{1}{2}\right)^{\frac{d}{8}} = e^{kd}$$

Step 2: Rewrite both sides of the equation so they each have an exponent of d.

$$\left(\frac{1}{2}\right)^{\frac{d}{8}} = e^{kd}$$

$$\left(\left(\frac{1}{2}\right)^{\frac{1}{8}}\right)^{d} = \left(e^{k}\right)^{d}$$

Step 3: Since both sides are raised to the same power, the expressions raised to that power must be equal.

$$\left(\frac{1}{2}\right)^{\frac{1}{8}} = e^{k}$$

Step 4: If there is an exponential equation of the form $e^{a} = b$, then $a = \ln b$.

$$k = \ln\left(\frac{1}{2}^{\frac{1}{8}}\right) \approx -0.087$$

The correct choice is **(1)**.

19. Create a table of values of the function $y = \log_2 x$:

x	$y = \log_2 x$
4	2
2	1
1	0
$\dfrac{1}{2}$	-1
$\dfrac{1}{4}$	-2

When these five points are plotted and joined, the graph looks like this:

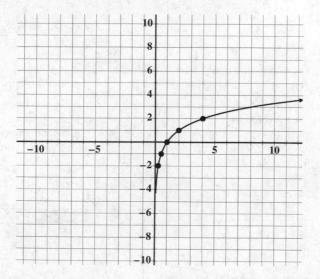

When the points are translated to the right 1 unit and 1 unit down, the graph looks like this:

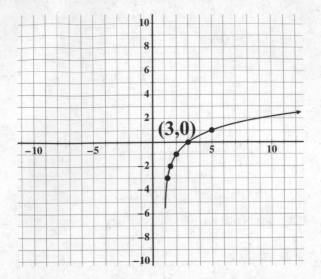

The point that originally had coordinates $(2, 1)$ now has coordinates $(3, 0)$. Therefore, $(3, 0)$ is the x-intercept of the translated graph.

Another way to solve this is to create the equation for the translated curve. Since the graph of $y = \log_2 x$ is translated 1 unit to the right and 1 unit down, the new equation is $y = \log_2 (x - 1) - 1$. If you graph this new equation on a graphing calculator and use the zero feature, you can find the x-intercept.

For the TI-84:

For the TI-Nspire:

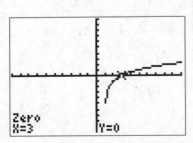

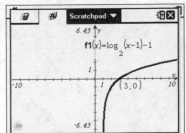

The correct choice is **(4)**.

20. Simplify the two expressions, and combine like terms. A rule of exponents that gets used several times when solving this problem is $x^{\frac{n}{m}} = \sqrt[m]{x^n}$:

$$\sqrt{16x^2} \cdot x^{\frac{2}{3}} + \sqrt[3]{8x^5}$$

$$= \sqrt{16} \cdot \sqrt{x^2} \cdot x^{\frac{2}{3}} + \sqrt[3]{8} \cdot \sqrt[3]{x^5}$$

$$= 4 \cdot x \cdot x^{\frac{2}{3}} + 2 \cdot x^{\frac{5}{3}}$$

$$= 4x^{\frac{5}{3}} + 2x^{\frac{5}{3}}$$

$$= 6x^{\frac{5}{3}}$$

$$= 6\sqrt[3]{x^5}$$

The correct choice is (2).

21. When the directrix of a parabola is a horizontal line and the focus is below the directrix, the parabola opens downward. When written in vertex form, the equation is $(x - h)^2 = 4p(y - k)$, where the vertex is (h, k) and p is the y-coordinate of the focus minus the y-coordinate of the vertex.

The following is a graph of the directrix, focus, and the parabola defined by that focus and directrix:

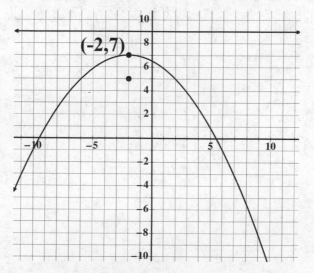

Since the vertex is halfway between the focus and the directrix, the vertex is $(-2, 7)$:

$$p = 5 - 7 = -2$$

When written in vertex form, the equation of this parabola is:

$$(x - (-2))^2 = 4 \cdot (-2)(y - 7)$$
$$(x + 2)^2 = -8(y - 7)$$

The correct choice is **(4)**.

22. Simplify the first expression:

$$x = (a + b + c)^2$$
$$x = (a + b + c)(a + b + c)$$
$$x = (a + b + c)a + (a + b + c)b + (a + b + c)c$$
$$x = a^2 + ab + ac + ab + b^2 + bc + ac + bc + c^2$$
$$x = a^2 + 2ab + 2ac + b^2 + 2bc + c^2$$
$$x = a^2 + b^2 + c^2 + 2ab + 2bc + 2ac$$
$$x = a^2 + b^2 + c^2 + 2(ab + bc + ac)$$

It is also given that $y = a^2 + b^2 + c^2$ and $z = ab + bc + ac$.

Substituting y and z into the expression $x = a^2 + b^2 + c^2 + 2(ab + bc + ac)$ gives $x = y + 2z$.

The correct choice is **(4)**.

23. Use the law of exponents $x^{ab} = (x^a)^b$ to check which of the four choices is equivalent to the given expression:

Choice (1):

$$29,400\left(1.068^{\frac{1}{365}}\right)^t \approx 29,400(1.00018)^t \neq 29,400(1.068)^t$$

Choice (2):

$$29,400\left(\frac{1.068}{365}\right)^{365t} \approx 29,400(0.002926)^{365t}$$

$$\approx 29,400\left(0.002926^{365}\right)^t$$

$$\approx 29,400(0)^t \neq 29,400(1.068)^t$$

Choice (3):

$$29,400\left(1+\frac{0.068}{365}\right)^t \approx 29,400(1.0001863)^t \neq 29,400(1.068)^t$$

Choice (4):

$$29,400\left(1.068^{\frac{1}{365}}\right)^{365t} = 29,400\left(1.068^{\frac{1}{365}\cdot 365}\right)^t$$

$$= 29,400\left(1.068^1\right)^t = 29,400(1.068)^t$$

The correct choice is **(4)**.

24. Since y pounds is equal to $16y$ ounces, the weight of each gift set, in ounces, is $x + 16y$. To find how many of these gift sets hold 124 ounces, divide 124 by the weight of one gift set, $\frac{124}{x+16y}$.

The correct choice is **(3)**.

PART II

25. The key word in this question is the word *given*. Since it is given that the student prefers to watch reality shows, the denominator of the fraction is the total number of students who prefer reality shows. This is $110 + 103 = 213$. The numerator of the fraction is the number of freshmen who prefer reality shows, which is 103.

   The exact probability is $\dfrac{103}{213}$. Do not convert this to a decimal since it will no longer be an exact value.

26. Make a table of values.

x	y
−1	−10
0	6
1	10
2	8
3	6
4	10

   Plot these 6 ordered pairs on the graph, and draw the curve that passes through them. Since only the x-values between −1 and 4 are in the domain, make the points (−1, −10) and (4, 10) the endpoints of the curve. Do not make arrows showing that the curve could continue.

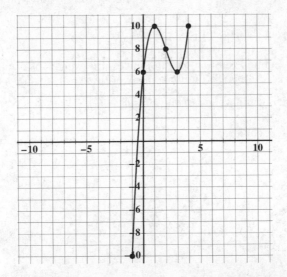

27. Use the quadratic formula with $a = 2, b = 5$, and $c = 8$:

$$x = \frac{-b \pm \sqrt{b^2 - 4ac}}{2a}$$

$$x = \frac{-5 \pm \sqrt{5^2 - 4 \cdot 2 \cdot 8}}{2 \cdot 2}$$

$$= \frac{-5 \pm \sqrt{25 - 64}}{4}$$

$$= \frac{-5 \pm \sqrt{-39}}{4}$$

$$= \frac{-5 \pm i\sqrt{39}}{4}$$

$$= -\frac{5}{4} \pm \frac{\sqrt{39}}{4} i$$

28. To avoid bias, a random sampling is required. In this scenario, the people who participate are self-selecting. Therefore, this data collection method is not a random sampling.

29.

$$2a^2 + 5a + 2 + \left(\frac{-5}{3a - 2}\right)$$

$$3a - 2 \overline{\smash{)}6a^3 + 11a^2 - 4a - 9}$$

$$\underline{-(6a^3 - 4a^2)}$$

$$15a^2 - 4a$$

$$\underline{-(15a^2 - 10a)}$$

$$6a - 9$$

$$\underline{-(6a - 4)}$$

$$-5$$

The final solution is $2a^2 + 5a + 2 - \dfrac{5}{3a - 2}$.

30.

$$a_1 = 3$$
$$a_2 = 1 + 2a_1 = 1 + 2 \cdot 3 = 7$$
$$a_3 = 1 + 2a_2 = 1 + 2 \cdot 7 = 15$$
$$a_4 = 1 + 2a_3 = 1 + 2 \cdot 15 = 31$$

This sequence can be represented as an explicit geometric formula only if the ratio of consecutive terms is always the same. Since $\frac{7}{3} \neq \frac{15}{7} \neq \frac{31}{15}$, this sequence cannot be represented as an explicit geometric formula.

31. The given information is the following:

$$P = 152,500 - 15,250 = 137,250$$
$$n = 30 \times 12 = 360$$
$$r = 0.036$$

Substitute these values into the formula:

$$M = \frac{137,250\left(\dfrac{0.036}{12}\right)\left(1 + \dfrac{0.036}{12}\right)^{360}}{\left(1 + \dfrac{0.036}{12}\right)^{360} - 1} \approx 624$$

Their monthly payment to the *nearest dollar* is $624.

32. Draw a circle with center (0, 0) and that passes through (2, −1).

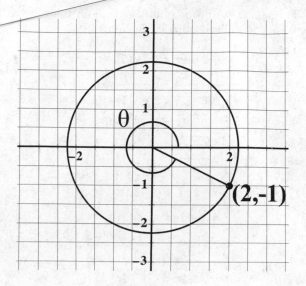

The radius of this circle is $r = \sqrt{2^2 + (-1)^2} = \sqrt{4+1} = \sqrt{5}$. The y-coordinate of the point is −1. Calculate the *exact* value of sine:

$$\sin \theta = \frac{y}{r} = \frac{-1}{\sqrt{5}}$$

PART III

33. Isolate the radical expression, and square both sides of the resulting equation. Then solve the quadratic equation:

$$\sqrt{6-2x} + x = 2(x+15) - 9$$

$$\underline{-x = -x}$$

$$\sqrt{6-2x} = 2(x+15) - 9 - x$$

$$\sqrt{6-2x} = 2x + 30 - 9 - x$$

$$\sqrt{6-2x} = x + 30 - 9$$

$$\sqrt{6-2x} = x + 21$$

$$\left(\sqrt{6-2x}\right)^2 = (x+21)^2$$

$$6 - 2x = x^2 + 42x + 441$$

$$\underline{-6 \quad\quad = \quad\quad\quad\quad -6}$$

$$-2x = x^2 + 42x + 435$$

$$\underline{+2x = \quad\quad\quad +2x}$$

$$0 = x^2 + 44x + 435$$

$$0 = (x+15)(x+29)$$

$$x + 15 = 0 \quad\quad\quad\quad\quad x + 29 = 0$$

$$\underline{-15 = -15} \quad \text{or} \quad \underline{-29 = -29}$$

$$x = -15 \quad\quad\quad\quad\quad\quad x = -29$$

A solution that does not satisfy the original equation needs to be rejected. Test both solutions to see if either, or both, of them need to be rejected.

Check $x = -15$:

$$\sqrt{6 - 2(-15)} + (-15) \stackrel{?}{=} 2(-15 + 15) - 9$$

$$\sqrt{36} - 15 \stackrel{?}{=} 2(0) - 9$$

$$-9 = -9$$

$$\checkmark$$

Check $x = -29$:

$$\sqrt{6 - 2(-29)} + (-29) \overset{?}{=} 2(-29 + 15) - 9$$

$$\sqrt{64} - 29 \overset{?}{=} 2(-14) - 9$$

$$8 - 29 \overset{?}{=} -28 - 9$$

$$-21 \neq -37$$

Since -29 is rejected, the only solution is -15.

34. The notation for the mean is $\bar{x}$. Calculate the difference in the means (scented − unscented):

$$23 - 18 = 5$$

To the *nearest hundredth*, the interval representing the middle 95% of the difference in means is between the following two values:

$$\text{lower} = \text{mean} - 2 \times \text{st. dev.}$$
$$\text{lower} = 0.030 - 2 \times 1.548$$
$$\text{lower} \approx -3.07$$

$$\text{upper} = \text{mean} + 2 \times \text{st. dev.}$$
$$\text{upper} = 0.030 + 2 \times 1.548$$
$$\text{upper} \approx 3.13$$

The difference of means in the experiment is 5, which is greater than 3.13. Therefore, the difference in means in Joseph's experiment is statistically significant.

35. The formula for compound interest is $A = P\left(1 + \dfrac{r}{n}\right)^{nt}$, where P is the starting amount, A is the ending amount, r is the interest rate, t is the time in years, and n is the number of compoundings each year. For this example, $A = C(t)$, $P = 63{,}000$, $r = 0.0255$, and $n = 12$:

$$C(t) = 63{,}000\left(1 + \frac{0.0255}{12}\right)^{12t}$$

To find the number of years, to the *nearest hundredth of a year*, for the account to reach $100,000, plug in $C(t) = 100,000$ and solve for t by using logarithms:

$$100,000 = 63,000\left(1 + \frac{0.0255}{12}\right)^{12t}$$

$$\frac{100,000}{63,000} = \frac{63,000\left(1 + \frac{0.0255}{12}\right)^{12t}}{63,000}$$

$$\frac{100}{63} = \left(1 + \frac{0.0255}{12}\right)^{12t}$$

$$\frac{100}{63} = 1.002125^{12t}$$

$$12t = \log_{1.002125}\left(\frac{100}{63}\right)$$

$$12t \approx 217.6594$$

$$\frac{12t}{12} \approx \frac{217.6594}{12}$$

$$t \approx 18.14$$

36. First put your calculator in radian mode. To calculate the average rate of change of $h(t)$ on the interval from a to b, use the formula for average rate of change:

$$\text{average rate of change} = \frac{h(b) - h(a)}{b - a}$$

$$= \frac{h(2) - h(1)}{2 - 1} = \frac{2 - 14}{2 - 1}$$

$$= -12$$

In cm/sec, the average rate of change of the piston's height on the interval $1 \leq t \leq 2$ is -12.

The simplest way to determine when $h(t) = 0$ between 1 and 5 is to graph the function and find the x-intercept of the graph. Again, be sure that the calculator is in radian mode.

For the TI-84:

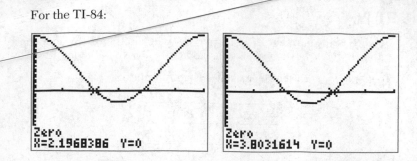

For the TI-Nspire:

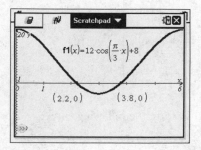

To the *nearest tenth of a second*, $h(t) = 2.2$ and $h(t) = 3.8$.

This could also have been calculated algebraically:

$$0 = 12 \cos\left(\frac{\pi}{3}t\right) + 8$$

$$-8 = 12 \cos\left(\frac{\pi}{3}t\right)$$

$$-\frac{8}{12} = \cos\left(\frac{\pi}{3}t\right)$$

$$-\frac{2}{3} = \cos\left(\frac{\pi}{3}t\right)$$

$$\frac{\pi}{3}t = \pi - \cos^{-1}\left(\frac{2}{3}\right) \approx 2.3 \qquad\qquad \frac{\pi}{3}t = \pi + \cos^{-1}\left(\frac{2}{3}\right) \approx 4.0$$

$$\text{or}$$

$$t \approx 2.3 \cdot \frac{3}{\pi} \approx 2.2 \qquad\qquad\qquad\qquad t \approx 4.0 \cdot \frac{3}{\pi} \approx 3.8$$

PART IV

37. Since x is the number of visits per week in thousands, the popularity rating, to the *nearest tenth*, is equal to $P(16) = \log(16 - 4) = \log 12 \approx 1.1$.

To graph $y = P(x)$, start by making a chart of values:

x	$P(x)$
6	0.3
8	0.6
10	0.8
12	0.9
14	1.0
16	1.07
18	1.14
20	1.2

When these points are graphed, they form the following curve:

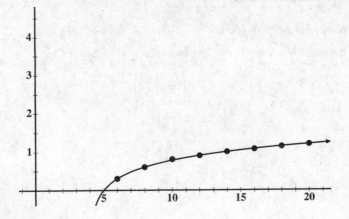

Make a chart for the same *x*-values to create a graph of $R(x)$:

x	$R(x)$
6	–3
8	–2
10	–1
12	0
14	1
16	2
18	3
20	4

When the graph of $R(x)$ is put on the same set of axes as the graph of $P(x)$, the graphs look like the following:

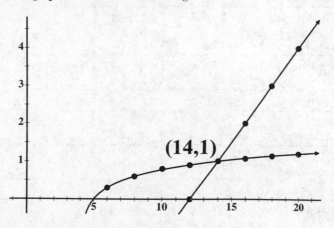

These curves intersect at the point (14, 1). This means that the two models provide the same rating, 1, for 14,000 weekly visits.

Topic	Question Numbers	Number of Points	Your Points	Your Percentage
1. Polynomial Expressions and Equations	1, 8, 12, 13, 21, 22, 29	2 + 2 + 2 + 2 + 2 + 2 + 2 = 14		
2. Complex Numbers	5, 27	2 + 2 = 4		
3. Exponential Expressions and Equations	2, 4, 7, 14, 18, 19, 23, 31, 35	2 + 2 + 2 + 2 + 2 + 2 + 2 + 2 + 4 = 20		
4. Rational Expressions and Equations	9, 24	2 + 2 = 4		
5. Radical Expressions and Equations	3, 20, 33	2 + 2 + 4 = 8		
6. Trigonometric Expressions and Equations	10, 32, 36	2 + 2 + 4 = 8		
7. Graphing	16, 26, 37	2 + 2 + 4 = 8		
8. Functions	6, 15	2 + 2 = 4		
9. Systems of Equations				
10. Sequences and Series	30	2		
11. Probability	11, 25	2 + 2 = 4		
12. Statistics	17, 28, 34	2 + 2 + 4 = 8		

HOW TO CONVERT YOUR RAW SCORE TO YOUR ALGEBRA II REGENTS EXAMINATION SCORE

The accompanying conversion chart must be used to determine your final score on the June 2018 Regents Examination in Algebra II. To find your final exam score, locate in the column labeled "Raw Score" the total number of points you scored out of a possible 86 points. Since partial credit is allowed in Parts II, III, and IV of the test, you may need to approximate the credit you would receive for a solution that is not completely correct. Then locate in the adjacent column to the right the scale score that corresponds to your raw score. The scale score is your final Algebra II Regents Examination score.

Regents Examination in Algebra II—June 2018
Chart for Converting Total Test Raw Scores to Final
Examination Scores (Scaled Scores)

Raw Score	Scale Score	Performance Level	Raw Score	Scale Score	Performance Level	Raw Score	Scale Score	Performance Level
86	100	5	57	82	4	28	67	3
85	99	5	56	81	4	27	66	3
84	98	5	55	81	4	26	65	3
83	97	5	54	81	4	25	64	2
82	96	5	53	80	4	24	63	2
81	95	5	52	80	4	23	61	2
80	95	5	51	79	4	22	60	2
79	94	5	50	79	4	21	59	2
78	93	5	49	79	4	20	56	2
77	92	5	48	78	4	19	55	2
76	92	5	47	78	4	18	53	1
75	91	5	46	78	4	17	52	1
74	90	5	45	77	3	16	50	1
73	90	5	44	77	3	15	48	1
72	89	5	43	76	3	14	46	1
71	88	5	42	76	3	13	44	1
70	88	5	41	76	3	12	42	1
69	87	5	40	75	3	11	39	1
68	87	5	39	75	3	10	36	1
67	86	5	38	74	3	9	34	1
66	86	5	37	74	3	8	31	1
65	86	5	36	73	3	7	28	1
64	85	5	35	72	3	6	24	1
63	84	4	34	72	3	5	21	1
62	84	4	33	71	3	4	17	1
61	83	4	32	70	3	3	13	1
60	83	4	31	69	3	2	9	1
59	82	4	30	69	3	1	5	1
58	82	4	29	68	3	0	0	1